Protecting Your Internet Identity

Protecting Your Internet Identity

Are You Naked Online?

Updated Edition

Ted Claypoole and Theresa Payton

ROWMAN & LITTLEFIELD
Lanham • Boulder • New York • London

Published by Rowman & Littlefield
A wholly owned subsidiary of The Rowman & Littlefield Publishing Group, Inc.
4501 Forbes Boulevard, Suite 200, Lanham, Maryland 20706
www.rowman.com

Unit A, Whitacre Mews, 26-34 Stannary Street, London SE11 4AB, United Kingdom

Distributed by NATIONAL BOOK NETWORK

British Library Cataloguing in Publication Information Available

Library of Congress Cataloging-in-Publication Data

Names: Claypoole, Ted, 1963– author. | Payton, Theresa, 1966– author.
Title: Protecting your internet identity : are you naked online? / Ted Claypoole and
   Theresa Payton.
Description: Updated Edition. | Lanham : Rowman & Littlefield, 2017. | Revised
   edition of the authors' Protecting your internet identity, 2012. | Includes
   bibliographical references and index.
Identifiers: LCCN 2016014745 (print) | LCCN 2016017408 (ebook) | ISBN
   9781442265394 (pbk. : alk. paper) | ISBN 9781442265400 (electronic)
Subjects: LCSH: Online identities—Social aspects. | Online identity theft—Prevention. |
   Internet—Social aspects.
Classification: LCC HM851 .C57 2017 (print) | LCC HM851 (ebook) | DDC
   302.3—dc23
LC record available at https://lccn.loc.gov/2016014745

⊗™ The paper used in this publication meets the minimum requirements of American National Standard for Information Sciences—Permanence of Paper for Printed Library Materials, ANSI/NISO Z39.48-1992.

Printed in the United States of America

CONTENTS

FOREWORD

Chris Swecker

On the Internet, governments, big business, private citizens, and criminals have all learned how to harvest and use personal data for many purposes—both legal and illegal. In fact, any person or business that cares to harvest that information and has developed a basic level of skill can do so. This is the main thrust of Theresa Payton and Ted Claypoole's excellent work, *Protecting Your Internet Identity*. They point out that although the Internet and Web represent some of the greatest technological innovations in the world, they present risks and dangers few Internet users appreciate. As a result, people fail to protect themselves from those who would exploit that information at the expense of safety, privacy, and even financial security.

Finally, there is a guide written by cyberexperts, not for technogeeks, but for the average Internet user. Cyberauthorities Payton and Claypoole explain in plain language how the World Wide Web is actually the "Wild Wild Web." They explain why we must open our eyes to the peril we are exposed to when we engage in routine activities such as opening a browser, accessing our e-mail, or paying our bills online. This book is required reading for Internet users because it simplifies critical concepts about the cyberenvironment and provides the reader with essential knowledge and tips on how to mitigate the dangers and become the master of your Internet persona.

The Internet is one of the last frontiers. It is barely regulated and never policed. When you access the Internet, there are no rules, and therefore no rules to enforce. As coauthor and Internet law expert Ted Claypoole points out, privacy laws are impotent when it comes to Internet-related privacy breeches, and there are only a handful of practical remedies. The book effectively paints the picture in terms we all can understand. We seldom stop to total how much sensitive information about ourselves we voluntarily consign to others in exchange for social interaction, a discount, or simply to access a product or service. This information can be our most private thoughts expressed on Facebook, purchases made while displaying our preferred customer card, our physical location via the

GPS on our mobile device, and even our financial data courtesy of our favorite financial institution. Inevitably this information ends up on the Internet, where it is vulnerable to being bought and sold by various businesses and marketing firms or stolen and exploited by tech-savvy criminal organizations.

The irony is not so much that we give the information voluntarily but that most of us have no idea how to exercise control over how that information is acquired and used. Theresa Payton is an authority on this subject, having held an executive-level technology security position at one of the world's largest financial institutions and worked on the front lines of the cyberwars as the chief information officer (CIO) for the White House. She and Claypoole present a tutorial on how we can control and effectively harness the information we expose for our own purposes, such as facilitating a business marketing plan or just to protect our privacy in a digital world. This is valuable information for people who are uneasy about exposing their information on the Web.

Chapter 6 describes the unlimited opportunities for cybercriminals to steal via the Internet. Theft of data is the perfect crime. It can be stolen from a computer in Russia, Bulgaria, or Romania, but unlike a car, jewelry, or a tangible object, it is not "missing." It's still there on your computer, and you don't notice something bad has happened until it's too late. As an FBI special agent for twenty-five years and ultimately the head of all FBI criminal investigations, I developed an acute understanding of how the Internet evolved to become the nesting ground and launching pad for the most sophisticated criminals in the world. The old brick-and-mortar crime model is outdated. In this new crime paradigm, the old adage that you can steal more money with a pen than a gun needs updating: you can steal more money with a computer than a gun. Cyberthieves never have to set foot in this country, making it difficult to investigate, and even more difficult to prosecute, violators.

Claypoole and Payton explain how the new black market currency is "personally identifiable information" (PII) and how these cybergangs use social engineering techniques such as phishing, pharming, whaling, and malware of every description to steal your user ID, password, or other sensitive information. Chapter 6 describes how this information is sold on the cyber black market and ultimately used to take over your bank accounts or even your identity.

Chapter 10, which deals with child predators on the Internet, is a must-read for parents with children who surf the Web, e-mail, tweet, Facebook, text, or routinely touch the Internet in any fashion. This chapter describes the dangers presented by pedophiles and sex offenders who troll the Internet for lonely teens and attempt to gain their trust. The ultimate goal of many of these deviants is to make personal contact with these vulnerable children for the purpose of sexual exploitation. It's not a pretty picture, but it is entirely preventable. This chapter alone is worth the price of the book.

Nothing that touches the Internet is secure. This has been widely acknowledged by U.S. government officials such as Gordon Snow, assistant director of the FBI's Cyber Crime Division, in his statement before the U.S. Senate Judiciary Committee on April 12, 2011, where he testified that "a determined adversary will likely be able to penetrate any system that is accessible directly from the Internet." The Internet is a high-crime neighborhood and must be respected if you are going to expose your personal information to every other human being on the planet.

If one were to prioritize chapters of this book in order of importance and relevance to your online wellness, I highly recommend dwelling on chapters 8 and 11, which deal with exercising your choice to be invisible online when you feel the need to do so. As the authors point out, our online anonymity and privacy are distinct from each other. Privacy is generally a legal standard, whereas choosing to be anonymous requires taking steps to disguise your true identity, which Internet users may do for valid reasons. This chapter is an invaluable aid to those doing Web research, blogging, or who are active social networkers. Following the advice provided is another step to clothing yourself and your loved ones from 800 million prying eyes. In this area knowledge is power.

This book is direct, digestible, and practical. Unfortunately, most works that deal with cybersecurity and data privacy are readable only by techies and attorneys who specialize in this area of the law. Most people know how to use the Internet and the latest electronic communication devices, but they are not interested in mastering the inner workings of the technology. Use of industry jargon and dissecting the technology behind firewalls and viruses or parsing complex privacy laws is like telling someone how to build a watch when they only need to know the time. Internet users don't need the subject further obscured or complicated; they need the same commonsense awareness levels of the risks and dangers that they have concerning their physical security, their houses, their cars, and their belongings. The most effective police anticrime campaigns don't dissect the laws and the technology behind burglar alarms, locking devices, or pepper spray; they arm you with sensible information and tips on how to avoid becoming an easy victim. Bravo for the authors, Claypoole and Payton, who have accomplished this with *Protecting Your Internet Identity*. This book is long overdue and will arm you with all the tools and knowledge you need to avoid risky, unnecessary exposure. Ignore their advice at your own peril.

CHAPTER I
HOW WERE YOU EXPOSED?

We are all born naked.

We emerge into this world with nothing to hide. But we are born into a complex human society, and it soon forces us to cloak ourselves in secrets. We choose to hide many aspects of ourselves from the world. Finances and romances, opinions and frustrations, imperfections and bad habits are all sensitive, personal information. The longer our lives, the more private information we accumulate.

Today the Internet threatens to strip us bare. By broadcasting many of our most sensitive and important secrets and keeping that information available and searchable indefinitely, the Internet displays aspects of our lives that we thought we'd kept private. Even worse, the Internet allows other people to collect facts about us and to aggregate those facts into a picture of our lives. The news is filled with stories about young people and celebrities who "tweet" their lives away, broadcasting their most intimate thoughts, feelings, and circumstances to anyone who will pay attention. The current world of reality television is built on the relationships between exhibitionists who will do anything for fame and voyeurs who find their actions fascinating. Social media sites such as Facebook and Instagram rely on their users' eagerness to share information—both intimate and mundane—in real time. Current culture is a fact-sharing machine, and the Internet is one of its most prominent engines.

This book starts with the assumption that some aspects of our lives should not be shared with everyone in the world and that you should have control over what you share and how you share it. We believe that privacy has value. Privacy protects our families and our peace of mind. Privacy is a strategy for shielding resources from thieves and our children from predators, it is a prudent business tactic for negotiations, and it is an important social tool when meeting new people. In this chapter, we look at how your personal information has become a commodity and just who is exposing you online.

1

Putting the Persona in Perspective

Elia Kazan, the great American film director and cofounder of the Actors Studio, said, "I am many men, but none of them is me." The various aspects of a personality can add up to a whole person, but no single aspect accurately portrays a person's life. The Internet persona is, in many ways, a separate self. This public self may reflect a portion of the private self, but the private self is always much more than the online persona reveals.

The Internet persona is a shadow of celebrity. You can learn more about your neighbors or some of your children's teachers than the most avid tabloid readers could have discovered about their favorite celebrities thirty years ago. When a private citizen tweets about her breakfasts on Twitter, shows her travel photos on Instagram or Tumblr, discusses her favorite team in public sports comments, and explains her business on LinkedIn and her company website, you could spend all day just following her movements. An Internet persona can be celebrated or excoriated, but it is most often ignored until a need arises to know more. So it rests and grows online, waiting for someone to care enough to pay attention.

The Historical Persona

Most of our parents and grandparents did not make distinctions between their public and private personas because they were known by their neighbors, family, and friends and no one else. Without the self-publishing tool of the Internet, private individuals remained private to the world and left only a trail of official notices in their wakes—birth records, wedding announcements, land transfers, and obituaries describing as much of their lives as their children wanted to expose. Before the Internet, a public persona was not an option or a problem to be managed for most people. It simply did not exist.

Of course, some people in past centuries who were published authors or entertainers led more public lives. Their lives provide an interesting example of the management and manipulation of public faces. These writers, politicians, and entertainers of the past help us understand how a public image differs from a private image. In some cases, a leader's supporters created myths to emphasize the leader's more admirable traits. Young George Washington never chopped down a cherry tree or said, "I cannot tell a lie," and a three-year-old Davy Crockett clearly did not kill a bear. The public persona of each celebrity was exaggerated for effect.

Sometimes writers created a different public persona to hide behind when writing dangerously controversial material. In the eighteenth century, François-Marie Arouet published his work under dozens of pseudonyms, including the name Voltaire. His influential writings on politics and the rights of man were inflammatory enough to earn him exile from his homeland. Many of the most influential arguments for the ratification of the U.S. Constitution were published in the *Federalist Papers* under the pseudonym Publius, and these were probably written by American founding fathers Alexander Hamilton, James Madison, and John Jay. These authors chose to develop public images that differed from their private lives.

The Internet Persona

In the age of the Internet, a public persona is forced on us, growing with or without our conscious contribution. Thanks to the Internet, we are all entertainers and publishers now. We can all send thoughts, opinions, and videos of ourselves throughout the world with the click of a mouse or tap of a finger. Hundreds of millions of people have Facebook pages, LiveJournal diaries, Flickr picture archives, or postings on other social media sites. Millions of people comment on public message boards for the *New York Times* online, CNET, the Fox News website, and ESPN.com. Thousands of new blogs are published every day. Now that everyone is able to entertain or publish online, the Internet persona is a fact of life for all of us, and it is a permanent, written record of our lives out there for everyone to see.

The Growth of Your Online Personas

Information about each of us collects on the Internet. This happens whether we approve of the information and whether we intentionally contribute it. As you'll see later in this chapter, you and others are building your online persona through a variety of activities. The things you write about yourself and your life, the pictures you or others upload, define you for many Internet users.

Even people who never go online have information about them posted on the Internet. Significant information about your life is available from public databases. Organizations that you join may provide facts about you on the Internet. Friends may expose information and never think to ask you if that's okay. Coworkers may post information about you on their social media pages.

This set of information may not be an accurate description of you, but because it's easy to find on search engines, this is what many people consider to be the truth about you. You can ignore your online persona and let it grow unchecked, or you can measure it and manage it, just as famous entertainers or authors manage their stage or literary personas.

One Example: Online Scandal

Consider a scandal in Washington, D.C., played out online but with consequences in the real world.

Jessica Cutler, twenty-six years old, worked as an aide to Ohio senator Mike DeWine. After work, Cutler led an active social life, but when she decided to document her social liaisons in the nation's capital, her life took a wrong turn that ended badly for her.[1]

Cutler created an Internet diary, called a *blog*. Her blog was anonymous and published online under the title of *Washingtonienne*. The *Washingtonienne* blog created a scandal as readers tried to guess the identities of the writer and her paramours. She described frequent sexual liaisons with men in her life, writing at one point that she was currently having sex "with six guys. Ewww."

It's easy to see why the *Washingtonienne* blog became required reading for so many people working in D.C. Nearly every day brought news of another sexual rendezvous, including the Washington hangouts where meetings occurred, intimate descriptions of what happened, and the writer's evaluation of her feelings about the men involved and about her own behavior. She discussed her lovers' high-powered political jobs, but she protected their identities with a mysterious letter code. No one knew who the *Washingtonienne* was or who she was meeting. Her blog made it seem that she could be sitting next to you at a Georgetown bar or an Arlington restaurant on any given night, then going home or to a hotel for outrageous carnal activity, only to jump online the next morning and tell everyone about it.

She claimed to be trading sex for money with powerful men, writing, "Most of my living expenses are thankfully subsidized by a few generous older gentlemen. I'm sure I am not the only one who makes money on the side this way: how can anybody live on $25K/year??"

Anonymous Internet writers had created hoaxes before and the Washingtonienne's stories seemed too lurid to be true, yet the details seemed too specific to be a deception. People talked about her in their offices. Who was the Washingtonienne, and did she really work on Capitol Hill? How was she juggling this many relationships? Was it true that a presidentially appointed chief of staff was paying her for sex?

Her life, which seemed so out of control to readers of her blog, finally crashed. The Washingtonienne was fired from her job on Senator DeWine's staff for misuse of government computers. This was the last post before Washingtonienne's firing: "I just took a long lunch with X and made a quick $400. When I returned to the office, I heard that my boss was asking about my whereabouts. Loser."

Another female Washington, D.C., government blogger, Ana Marie Cox of the popular policy blog *Wonkette*, named Jessica Cutler as the author of the *Washingtonienne* blog. Ms. Cox ran an interview with Ms. Cutler on the *Wonkette* blog, and the *Washington Post* soon followed suit with a full-feature story including pictures of the mysterious Washingtonienne.

Ms. Cutler's secret identity as the Washingtonienne affected her life in many ways, apart from the lost job in the U.S. Senate offices. Predictably, Ms. Cutler posed naked for *Playboy* and was offered a book deal worth a reported $300,000 advance. Her book inspired a *Washingtonienne*-based television series produced by HBO. She was also sued by one of her coworkers, who alleged that he was discussed in the *Washingtonienne* blog as one of her many lovers. Ms. Cutler ended up filing for bankruptcy.[2]

Cutler was literally and figuratively naked online. She developed an online persona, and it took over her life. She believed she could hide behind an anonymous Internet pen name, but in the end, her online persona merged with her real life of work, family, and friends. She was not the first to develop a separate online persona or the first to make money from doing so. Bloggers with online pseudonyms like Perez Hilton, the Daily Kos, and Lonelygirl15 boast millions of readers.

Although writing salacious autobiography has long been a path to celebrity, today's Internet provides fame and infamy to people who are clever or even unlucky with a smartphone camera. For example, more than forty million viewers subscribe to a YouTube channel where they watch a young Swedish man play video games and comment on the action. The man calls himself PewDiePie, and he made more than $7 million in 2014 from his YouTube following.[3] Another video channel called Vine only allows clips of six seconds or shorter and has spawned movie deals and significant incomes to the people who produce and star in these tiny films. A company may pay up to $50,000 for a Vine star to use its product in the six-second video.[4] Paul Vasquez, a California firefighter and trucker, became an overnight Internet sensation when he was caught on video effusing over a double rainbow. He later turned his Internet fame into a sponsorship deal with Microsoft. He is known to the millions of people who have seen his video as "the double-rainbow guy."

The Moral?

We are all complicated people with many aspects to our lives, and we change our identities as we grow in life. Today's wild child is tomorrow's suburban housewife. Today's poor college student may be running a huge corporation tomorrow. Seeing one aspect of someone's life through the prism of Internet writing may provide insight into that person, but it displays a skewed and inaccurate overall portrait. Cutler may have matured into a sedate wife and mother, but many people will know her primarily for the wildness of her young, single years and the scandal it caused. Vasquez will eternally be tied to one excitable moment caught on video. An Internet persona can be dangerous for many reasons, but it can be particularly dangerous as a brief snapshot from which people draw broader conclusions for years to come.

How Information Is Treated Online

In results that surprised even the researchers, a study conducted by social scientists at the University of California–Berkeley and the University of Pennsylvania found that American adults between the ages of eighteen and twenty-four claim to care as much about online privacy as older adults.[5] The study also found that young people tended to not understand the laws concerning privacy protection and to overestimate how much legal privacy protection individuals receive online. The researchers determined that "young-adult Americans have an aspiration for increased privacy even while they participate in an online reality that is optimized to increase their revelation of personal data." And they're not alone.

Social media rewards its users for publishing more and more information all the time. Comments, picture postings, likes, and the use of popular hashtags build a person's popularity and raise the profile of the poster. Find and post the most and best beach photographs or architectural drawings on Instagram and you are more likely to find other people interested in the same things you are. Tweet the most about a political position and garner the attention of the top people promoting the same position. On social media, more information means more interaction and more chances at a meaningful connection to other people.

What Information Can Be Discovered Online?

Imagine that new neighbors are coming to live next door. You haven't seen them yet but know that they closed on the house and are moving in next month. What could you discover about them using the Internet, even if you don't know their names?

You know their new address, so you can find their names online in the county records describing the real estate transfer of their new house, and you can probably find out their current address. You can also discover how much the new neighbors paid for the house and what lender they used, if they have a mortgage.

If both names of a married couple are included in the housing records, you can search the wedding announcements in the archives of their hometown newspaper to get more information about them. For example, the *New York Times* provides a searchable archive of wedding announcements published since 1981, including full text and analysis of hundreds of thousands of nuptials.

So what could you learn from these sources? You can find out their full family names, their ages, their parents' names and occupations, where the couple attended school, and some financial information. A quick search for birth announcements related to the same couple may yield the names and ages of their children. Many towns also keep dog license registration records online, so you may be able to find out what breed they own and even the dog's name.

You can do all of this research on public Internet sites without ever running a general search for either neighbor's name using a search engine such as Google, Yahoo!, or Bing. But if you need more information to perform ID theft or stalk a family member, you'll find that general searches can unearth employment information, family and genealogy data, social media postings made by family members themselves, and much, much more.

Exposure Is Rewarded

Everyone participating on the Internet exists in a world geared toward encouraging exposure of personal data. Social media sites are built to reward the sharing of information. The more people know about you on Facebook, the more points of connection they find and the more "friends" you will attract.

Information Rules

Think about the basic information most Facebook users reveal, then measure how many classmates, former coworkers, fellow Labradoodle lovers, cybercreeps, and long-lost family members are attracted by these revelations. Facebook's marketing pitch generally includes the concept that "you get more out of it if you put more into it." Your active participation in these sites is a cycle of personal disclosure and social or financial rewards for your level of sharing.

People use social media as a confessional, a watercooler, and even a psycho-analyst's couch. They tell secrets, they cry out in anguish, and they beg for other people to react. In many Internet communities, the deepest secrets and rawest emotions are rewarded with the warmest words of acceptance. Author Dave Eggers, exploring how people use social media as a tool for human connection, wrote in his book *The Circle*, "Suffering is only suffering if it's done in silence, in solitude. Pain experienced in public, in view of loving millions, was no longer pain. It was communion."[6]

In addition, commercial websites, from newspapers to banks and stores selling goods and services online, can profit from knowledge of their customers and visitors. Those sites encourage visitor participation and often place software called *cookies* onto visitors' computers. Cookies allow the sites to recognize your computer when you visit, track your shopping activity, make suggestions of items you might like, and even greet you by name.

Cookies and other tracking technologies also allow owners of commercial sites to better understand the habits of their Internet visitors and often to sell that information to advertisers. Those advertisers can then create more targeted advertising.

For example, have you ever noticed that the same banner advertisements seem to follow your browser as you click through various Internet pages? Why does your spouse always see ads about sports cars, whereas you see ads about cooking? Some sites make assumptions about you and use these assumptions to place you in an advertising program based on your online activities and calculated to interest you. You would be surprised (and not a little frightened) at the information they collect about you from a variety of sources to make these assumptions.

Although capturing your data for online advertising has always been surreptitious and hard to spot, a new trend sneaks its signals right past your ears. As both Internet browsing and online marketing become more sophisticated, the tracking tools change. Following the rise of smartphones and tablets to access the Internet, advertisers were faced with a new challenge to know their prospective customers: how can a marketer know that the "John Lee" who accessed the Levi's website from his laptop computer is the same "John Lee" who accessed the Levi's website from his smartphone, or his tablet, or his television, or his car, or his desktop computer at work? Stitching together the knowledge of who is operating each of these devices is an exercise called *cross-platform marketing*, or, more specifically, *cross-device tracking* and attempts to learn about all of a user's access devices have grabbed the attention of the U.S. Federal Trade Commission. Marketers such as SilverPush have been experimenting with strange technology to track their customers across various Internet access devices.

Some advertisers are using subsonic signals—sounds that are too low or too high for people to hear—to force Internet access devices to communicate with each other. If you visit a SilverPush-enabled site on your laptop computer, the computer's speaker may send out one of these "silent" signals, and the microphone on your smartphone may recognize the signal and the smartphone may both store it, and find a way to signal back. SilverPush technology can also insert subsonic audio into television commercials that can be "heard" by your smartphone or computer without you even knowing the connection exists. This method uses device proximity in the real world to discover more information about you, rather than relying on your Internet activities.

What Amazon Knows about You

But online collection of your information is amazing too. Amazon is a major online retailer, selling everything from clothing to cookware to electronics. This company uses its knowledge of the books, music, and other products that you have searched for or bought, including how long you spent exploring any single topic or item, to suggest additional products that you might be interested in.

If you read descriptions of or buy several books on French cooking, for example, you may be shown other popular books on the topic for your consideration. If you buy a vintage Frank Sinatra album or the newest hit from Kendrick Lamar, Amazon will propose other music from the same artist or genre that customers have purchased recently. This site even encourages you to suggest music playlists or literary reading lists to guide shoppers who may share your preferences, but what they're really doing is collecting information about your interests.

Once again, on this Internet commerce site, just like social media sites, your information is solicited and providing more data is rewarded by the site. Not only does the site track your movements and purchases, but it also solicits your comments and opinions on a broad range of topics, from books you have read to the service provided by Amazon. You are encouraged to return to the site and offer reviews of any books, music, or other products that you purchased there. Your reviews are supposed to provide other shoppers with the benefit of your analysis, but at the same time, they give Amazon more information about you.

Amazon allows other product users to rank the helpfulness of your review, providing yet another reason for you to return and check how the community responded to your wisdom. Each of these acts of sharing is supposed to enrich your shopping experience at Amazon, to make the site's anticipation of your wishes more accurate, and to make you feel more like a member of a community.

For Amazon, enriching your online experience in these ways is a psychological technique to keep you in the website longer and draw you back to the online store more often. It is the Internet equivalent of providing a coffee bar and comfortable furniture in a brick-and-mortar bookstore to make you feel more at home and to encourage you to browse, read, and buy. However, in the online version of this strategy, you provide Amazon, and maybe other Amazon customers and partners, with a wealth of information about you and about your preferences. Amazon takes the information gathering a step further by offering its Amazon Prime service. For a relatively small annual payment, Amazon Prime opens a wide selection of benefits to the Prime customer, from free shipping for items purchased on the main Amazon website to a lending library for people who own Amazon's Kindle tablets. Prime ties an Internet user closer to Amazon, because once shipping fees are waived then shopping for nearly anything Amazon offers, from music to gardening tools, becomes easier and as cost effective as running to a local store. Membership in Prime also provides Amazon with more information about your tastes and priorities.

Amazon is not by any means alone in these practices, and, in unscrupulous hands, these same practices can be used for much more than selling you a book or DVD.

Why Now?

Why worry now about my online persona? The Internet has been with us for a long time—why have we not been reading about issues of privacy in the early years? Although the Internet has been available to the general public for more than twenty-five years now, the way that it works and the sharing of personal information have changed drastically over time. At first, the Internet was used for computer file transfers, electronic mail, and text-based chat groups. As browser software became popular and millions of people joined content-heavy services such as CompuServe, Prodigy, and America Online, they learned to find interesting information about government, businesses, or simply other people.

E-commerce began to flourish online in the 1990s, and within a decade, nearly every commercial and consumer business felt the need to supplement its sales with some kind of online store. The cost of computer storage dropped drastically in the early 2000s. In addition, the development of technologies such as video streaming and video sharing allowed websites to use more sophisticated graphics, video, and audio files.

The era of Web 2.0, with increased interactivity between Internet users and websites, brought with it the possibility that every user of the Internet could not only receive information but also share their information and interact with others. Applications accessed from devices such as tablets and smartphones have moved Internet usage to a mobile platform. And where the Internet was once reached primarily through devices created for the task, online access is rapidly evolving into an activity that can be enjoyed everywhere from every device—from automobiles to airplanes to television sets—connectivity is spreading to a vast array of machines. These machines also collect and manage information about their users.

All of these changes have created a separate realm, accessed by anyone with the right cell phone or computer, where people learn and share information about each other. It has only been within the last decade, with the rise of social networks and the avalanche of personal information migrating online, that most of us have developed a substantial online persona. And the issue is likely to continue growing in importance as the Internet expands its reach into our personal lives.

Now is the time to recognize that you have an online reputation and to take control of it before years of information accumulates.

Who's Looking at You Online?

Before you think about exactly how you might be exposing yourself online, consider who's looking at your information. Your friends are not the only people examining your Facebook page. A study conducted for Careerbuilder.com found that 45 percent of companies search social networks to screen employment candidates.[7] Your spouse's divorce lawyer is looking, too. A survey conducted by the American Academy of Matrimonial Lawyers showed that 66 percent of divorce lawyers claimed that Facebook was their primary source of evidence.[8] And your Facebook postings may affect your service on jury duty. For example, the district attorney's office of Cameron County, Texas, based in Brownsville, incorporates social media as part of its jury screening process.

Once information is out there and publicly accessible, it can be viewed by any individual or organization, and it will be used to draw conclusions about you

FEATURED WEBSITE: THE INTERNET ARCHIVE

When people tell you that information on the Internet lasts forever, they're right, largely because of the existence of the Internet Archive. The Internet Archive is a nonprofit organization, classified as a library in the state of California. The library supports an online film archive, one of the world's largest book digitization projects, technology for an online lending library, and a distributable digital media collection, including otherwise unavailable audio and video files. But the Internet Archive is perhaps best known for its capture and collection of historical records of website content.

Also known as the "Wayback Machine," the Internet Archive's website archiving service keeps searchable, linkable copies of Internet sites as those sites existed in the past. If you want to know the board members of your local symphony orchestra in 2004, search the orchestra's website in the Wayback Machine. Or search the archive if you want to check a friend's online biography posted by the company she worked for two years ago or read her review of shoes she bought on a retail site.

Hundreds of millions of sites are available for historical research and reference. Since 1996, the Wayback Machine has sent software crawling around the World Wide Web and snapping archive copies of various Internet sites from governments, businesses, and private citizens. The Wayback Machine only collects publicly available websites, not sites that require a password. Not every site is archived, and a site owner can ask to be excluded from the archive.

As of the publication of this book, you can find the Wayback Machine at www.archive.org. According to the Internet Archive site, the Wayback Machine currently includes twenty-three petabytes of data and is growing at a rate of twenty terabytes per month. (A petabyte is a unit of information equal to one quadrillion bytes of data, or 1,000,000,000,000,000 bytes.) The Internet Archive also includes a mirrored copy site in Alexandria, Egypt. Because of technical complexities, it can take six months to two years for recently collected websites to appear in search results on the Wayback Machine.

11

as a romantic partner, potential spouse, employee, church member, potential victim of a crime, or parent. (See chapter 2 for more about this topic.)

Who Is Exposing You?

With the right tools, the Internet allows each one of us to customize our own Internet presence. Every Internet user can be his or her own broadcaster, sending opinions and preferences out to the world. People create blogs and diaries online, spilling their deepest thoughts into the cosmos. Social networks provide a space for people to display information about themselves but also to display their networks of friends and preferences for everything from food to relationships.

Today you may be broadcasting your own information online, but you're not the only one contributing to your online persona. Organizations, governments, friends, family, and media are also out there exposing you every day. In this section we look at the various sources exposing you online.

You Did It Yourself

The Internet's function of self-publication has revolutionized the way that humans communicate with each other. If you don't believe that, spend a day with a teenager and see how she uses social media, text messaging, instant messaging, and Skype video communication to stay in touch with friends both near and far.

But like all revolutions, the Internet communication revolution includes a negative side. Anyone who participates fully in social media, blogs, and the opportunity to comment on favorite websites is revealing much to the world. The person exposing most of your personal information online is probably you.

The Internet sites and tools discussed in this chapter are not the only ways to display yourself on the Internet. The Web contains millions of places you can publish information about yourself using a variety of technologies, with more appearing every day. Not surprisingly, there are very few tools to teach you self-censorship.

By posting a picture on the Internet and identifying yourself, you have just provided information about your age, gender, race, your health, your social class, your self-esteem, and your tastes. You may have included an image of your home, a favorite vacation spot, your car, or family members or friends, revealing even more about your life. Videos that you post may only multiply the exposure.

One Example: Facebook

If you are naked online by exposing personal information to the world, then there is a strong probability that you have flashed the world with your Facebook page. As of this writing, Facebook is claiming more than one and a half billion current active users. A current active user is a person with a Facebook page who has visited the site within the past month. Given those numbers, there are more than twice as many Facebook users than the total populations of the United States, Mexico, and Canada combined.

What is this staggering number of people doing at a single Internet site? They are posting information about themselves and reading and responding to information posted by other people. Facebook continues to add new tools to help you provide more information about yourself to anyone interested in learning about you.

The growth of photo and video posting is also astronomical. Facebook claims that its current *daily* photograph uploads average 360 million.[9]

Facebook includes a place to write messages viewable by everyone, including messages to small groups and messages that can be seen by just one person. Hundreds of millions of conversations on Facebook happen out in the open for everyone to read.

Facebook can also help people locate you at any time. The service offers a tool for you to tell the system exactly where you are standing at that moment—at the grocery store, on vacation in Bali, attending the soccer game, or at home in your kitchen—so that all of your Facebook friends, or all 1.5 billion and growing Facebook users, depending on your privacy settings, can discover your physical location. Criminals can even use the collected location data to understand your daily routine—for example, when you leave your house for work or when you buy groceries each week. This ability to locate anyone may seem offensive or intriguing to you, but when you think of someone knowing your child's every move, it's a use of technology that becomes frightening.

The bottom line: If you choose to accept all of the offered Facebook invitations to share information, many of the important facts, routines, people, and passions in your life will be available to millions of people.

Expressing Yourself in Comments

Spotlighting your own life on social networks is not the only way that you expose yourself online. Many commercial Internet sites have comment features that allow visitors to post opinions and responses. You could be singing the praises of the most comfortable pair of shoes you have ever owned or looking for sympathy on a relationship site because you just broke up with your boyfriend.

Your postings may include Amazon book or music reviews, or they may be political statements on the Fox News website or on *Huffington Post*. Your responses may be on the website of your favorite sports team or the ESPN page discussing your team's greatest rival. You may say positive or negative things about a public company on message boards tied to the performance of that company's stock price, or you may be commenting on a friend's loud Hawaiian shirt in weekend party pictures.

Whatever you say and wherever you post your comments, you expose your opinions, ideas, and thought processes to billions of people. Many of these posts are made under pseudonyms or "handles" that are not easy to trace back to you. However, it is possible to decipher who owns a handle, and keep in mind that anyone who learns your handle for posting on a specific website can learn a great deal about how you think and information about your life. (We discuss later in this book how uncovering a handle on social websites may be much easier than you think.)

Finally, the mother of all online comments is the blog. Publishing your own blog is similar to writing an updated page on a social media site, except the blog tends to focus on an area of interest and provides detailed analysis of your thinking on the subject. Many blogs are updated every day, and most well-known blogs include at least two new entries a week. A constant stream of words on nearly any subject can tell the world about your thinking process and probably leave clues about your work or home life. A blog often is little more than a lengthy online comment that includes thousands of words of self-revelation.

At the beginning of this chapter we discussed the blog of the Washingtonienne. Her blog is not an isolated phenomenon. According to Wordpress's website, in 2016 more than 409 million people viewed more than 21.5 billion blog pages each month, and Wordpress bloggers produce about 55.8 million new posts each month. These are just a small fraction of online content that can be classified as blogging. These Internet publications could be the random musings of madmen, detailed discussion of politics or technological products, or the growing phenomenon of mommy blogging—providing tips, product reviews, and shopping deals for raising kids in a specific location.

Twitter is a tool for microblogging, blogs that have a limited number of characters of text per posting. The Twitter technology allows pictures and small text messages to be posted online and sent directly to anyone who "follows you" on Twitter. Whereas some people use Twitter to post details on their daily commute and every mundane thought that enters their brains, others use the technology to organize political rallies, to call attention to the everyday movements of celebrities, or to lead teams on scavenger hunts. Twitter may be intrusive and pervasive, revealing everything about the writer from his or her deepest thoughts to up-to-the-minute location data.

Finally, you could be exposing your personal habits to others based only on the type of websites you frequent. The Internet offers interest-specific sites for stamp collectors, rugby players, and people who suffer from rare skin diseases. The fact that you have chosen to spend your time or become a member of any of these sites can speak volumes about you. In 2015, the Ashley Madison website

ARE YOU EXPOSING YOURSELF?

Add the numbers of your answers to find your score.

The information you post online is
1. next to nothing.
2. only the most basic information.
3. professional and business data only.
4. professional and personal.
5. everything I think or do, in real time.

Do you publish a blog or online diary?
1. No
2. Yes, but it can only be accessed by a small group of friends with whom I am close.
3. Yes, but it can only be accessed by my "friends," many of whom I have not met in person.
4. Yes, it is a public blog, but I never write about my personal life.
5. Yes, I write about myself for everyone to see.

How often do you post on Facebook?
1. Never, I don't have a Facebook page.
2. Less than once a month
3. At least once a week
4. Every day
5. Many times a day

Your personal Web presence is best described as
1. a few words of text.
2. anonymous reviews and comments.
3. a short, personal biography.
4. biography, pictures, and video.
5. all of the above, plus postings.
6. I post naked pictures of myself online.

Scoring

4–7	Careful and protective
8–11	Just testing the waters
12–16	Unabashed Internet junkie
17–21	Baring everything

was hacked and the names of people who had paid accounts on the site was published. Many of those people were humiliated or suffered significant trouble in their personal lives because of their exposure as Ashley Madison community members. This is because it is a website created for a person looking to cheat on his spouse to meet a partner who also wants to cheat on her spouse. Self-identifying publicly as an Ashley Madison member was akin to an announcement of attempted infidelity or worse. Website associations can speak volumes about you.

The Circle Widens: How Others Expose You

Posting information about yourself is entirely within your own control, but much of the information about you on the Internet is put there by someone else. As more of the world's data moves to the Web, information about you is probably part of it, often appearing where you least expect it.

Your friends, rivals, family, teachers, employers, church, and other connections may post information about you that others can see. Clubs are proud that you are a member, and businesses are pleased that you are a customer. These and other organizations and individuals may promote their associations with you online.

Exposure by Friends

Often without intending to, friends and family may be giving away your personal information. Once again, Facebook offers an example of how even well-meaning friends can expose you online. Facebook, as well as dozens of photography sites, offers digital tools that allow a user to post a photograph and then to "tag" all of the people in the picture. Tagging provides not only people's names but also links from the picture to the Facebook pages of all the people in the photograph. Your Facebook page will include links to any pictures posted by others in which you are tagged. These pictures may be embarrassing if they are reviewed by your boss, or they may show potential thieves where you live or the license plate of your car. You have little control over the picture's availability online.

But social networking sites aren't the only place others may expose you. Your grandmother may post your family tree on a genealogy site, giving an ID thief your mother's maiden name (often used as a password verification question) and much more. Your friend may send an e-mail to others telling them of your surprise party Friday night, along with your address and phone number. Anyone who knows you can post your information for all to see.

Being tagged in your friend's pictures can be fun, but it may also show a different side of you than you want exposed to the general public. How easy will it be for a future employer to find the wild party pictures from your friend's pages?

How Organizations and Your Employer Expose You

Your name, description, and picture are also likely to appear online if you serve on the board of an organization or are an active member at your local place of worship or community theater. These articles or images often give your name, so anybody can find them in Bing and Google searches. For example, while the United Nations Foundation is understandably proud that media mogul Ted Turner and Queen Rania Al Abdullah of Jordan serve on their board, the

INFORMATION PERMANENCE

Once information gets online, however it got there, different online entities may archive it for a long time. Although everyone knows about using search engines racing over the Internet to find available information, not many people understand the breadth and depth of current archiving projects. For example, Archives.com has collected a database of information commonly used for genealogy research, from birth and death records to family history and immigration certificates. Genealogy is a growing industry, so many other sites, including Ancestry.com, Familysearch.org, and Tribalpages.com provide similar services.

Google is the most active archiving company. Google is involved in a famous project to create searchable archives of all books printed in the history of humankind. Google also archives the physical structures of the world. The famous "Satellite" function on Google Maps allows prospective thieves to see the entire layout of your street and your property. If they have mapped your town, the Google Street View project also allows anyone in the world to look directly at your house from the street, and then see a 360-degree view around your neighborhood—the same view that you would see if you drove down your street and looked all around.

Google has also created the largest online archive of Holocaust photographs, an archive of *Life* magazine pictures, collections of the Prado Museum in Madrid, and the New York Public Library's historical postcard collection. Aside from the collections of images, videos, and maps it collects from the current Internet, Google also displays searchable patent archives and high-resolution digital images of historical maps from the David Rumsey Collection. These many and varied sources illustrate the fact that Google and other companies are taking information that was once only available in paper form and making that information searchable online.

website for the foundation includes their pictures and biographies as well as those of eleven other board members.

This practice is common for nonprofit entities. Volunteer organizations you're involved with aren't the only ones exposing you. Check out your employer's or school's website and you may find more information posted there than you are comfortable with, from your biography or résumé to a note about your participation in an upcoming, out-of-town conference.

Your Own Government Is Stripping You Naked

The vast majority of housing records within the United States have always been public information available to anyone. In the 1970s, a person looking up real estate records would have had to physically travel to the county recorder's office for each county that contained property he wanted to learn about. Once there, he would be directed to a back room filled with dusty, thirty-pound plat books to find one set of information and an entirely different set of books to find other types of information. The process was slow, laborious, and difficult. Now many U.S. counties keep their property records online so that any researcher can quickly run multiple record review requests from the comfort of his office or living room, discovering anything from what you paid for your house to the amount of your mortgage and your yearly income.

In addition to property records, the trend on public government sites is to keep updated information about current enforcement of laws and regulations but to maintain older data and press releases as well. Birth and death records, real estate transactions, arrests, convictions and traffic violations, marriage records, and nearly every other brush with public officials is recorded and posted. Local, state, and federal governments all keep public records. Testimony before Congress is online, hearings before many federal executive commissions are on the Internet, and so are the comments stated before local school boards and zoning commissions. Government sites, from courts to administrative agencies, treat their Internet sites as historical records, so they keep most information indefinitely.

One of the most embarrassing databases to move online is the database containing the records and opinions of court cases. Court proceedings have always been official public records, but now researchers and acquaintances can quickly and easily find out about your bankruptcy proceedings, your dispute with a former business partner, and maybe even your divorce settlement. These records show us at our worst and most stressful moments, and thanks to the government's embrace of the Internet, they are becoming much simpler to search and explore.

And some records last forever online. The FBI offers a sex offender registry where anyone can look up their own neighborhood and find people convicted of sex crimes. But other crimes are publicized as well. The state of Virginia offers names of everyone who was in the care of the state Department of Corrections for any crime. You do not even need to be convicted of a crime for this trend to affect you. For example, several public records sites offer mugshot databases that expose people who have been arrested for crimes. Once your mugshot is available online, people may quickly make up their minds about your character.

Exposure by Media

Newspapers were created to disseminate the important information of the day, including local news. They also discuss human-interest events such as weddings, funerals, and professional milestones such as job changes and promotions. Since the founding of the United States through the mid-1990s, newspapers have been an important source of information, but the notoriety only lasted for a day and then was relegated to the basement archives of the newspaper's offices. Only an intrepid researcher could fight through the dust to find paper copies or microfiche of each day's events.

Now not only are entire newspaper archives online, stretching back over decades, but the information in these papers is also searchable. Today a con artist or private investigator need only list the topics or person he is seeking, and any relevant articles or photos appear on the screen. No travel, no dust, no guessing—just answers.

Newspapers have jumped online in a big way. For example, Tampabay.com, the online site for the *Saint Petersburg Times* in Florida, allows visitors to its website to search its own archives back as far as 1987. Any news from West Central Florida for more than twenty years can be searched and displayed. If you were arrested in that town on spring break ten years ago, your ignominy will live forever in easily searchable newspaper text.

Aggregation search sites like Google go much further. In a single search, Google allows the researcher to look for topics covered by hundreds of newspapers back as far as 1901. Then the user can sort the results by relevance or by date. Google continues to add more newspaper and publication archives to this search function so that anyone can find news about you or quotes attributed to you from within news stories in newspapers throughout the world.

Of course, newspapers aren't the only media you can search online. If you appear on local or national TV or radio, that appearance could be searchable online. Strictly online news sources such as *Huffington Post* and MSN are also searchable. Whatever the media, its content is often indexed or archived somewhere on the Internet.

The Internet has spawned new types of media that can also add to your online persona. There is an entire category of websites that take pictures at parties, bars, and nightclubs and post them so we know who is attending the hot spots. Another cottage industry on the Internet consists of sites that publish photographs of celebrities and everyone seen with them. Like the newspaper society pages before them, these sites often identify the people in their pictures. The photographs and text can be searched, so you could be adding information to your online persona by attending the symphony fund-raiser or the hottest dance club in town and posing for pictures.

Now That You Know You're Naked, How Do You Get Dressed?

The Internet facilitates the greatest collection of information that has ever existed, and it is pushing deeper into data about our private homes, businesses, families, and lives. Sometimes we add to this information stream ourselves, and sometimes others enter information about us, but either way, the list grows longer and easier to access.

In the next chapter, we examine who is searching for information about you and what they want to know. Later chapters in this book address how to manage your online persona and constructively correct misleading or embarrassing information marring your Internet image. As you continue reading, you will learn what information affects you, your family, and your business, and what you can do to take control of your online data and reputation.

CHAPTER 2

PEEKERS AND GAWKERS
Who's Looking at Your Online Persona?

In May 2013 when the media reported that Edward Snowden had revealed information regarding digital surveillance programs around the world, citizens everywhere began to wonder who was looking at them and exactly what could they see.

We all have secrets or personal experiences we keep to ourselves or may share with a trusted circle of confidants but would be mortified if those secrets were exposed to the world. This begs the question, who would want to expose our private information, and what's in it for them? We know that third-party marketing firms are collecting every click and every piece of information we provide online. We also have resigned ourselves to the fact that our neighbors, friends, lovers, and prospective employers are going to "Google" us. Each company, organization, and person searching the Web may capture, preserve, and display facts about our lives that we might never reveal in a face-to-face conversation. In this chapter we run through the inventory of the types of individuals and organizations interested in you and how they might use your information in ways both legal and illegal.

Surveillance Center Stage

As the Edward Snowden story unfolded, many came to realize that some of the data involved were collected by private sector companies. Over the past fifteen years, although many countries have different definitions of what "privacy" means for their citizens, they have mostly agreed to disagree. The confluence of the Snowden revelations coupled with a lawsuit by an Austrian student, and the continued breaches of consumer data have led governments around the world to redefine what privacy means. They have also demanded greater accountability and protections for those companies that collect data about their citizens.

The European Court of Justice, considered the highest court in Europe, has canceled an agreement dating back to the year 2000. That accord permitted U.S. technology companies to collect information and allowed for digital data flow outside the geography of Europe using voluntary frameworks and rules on privacy. The United States and European Union are working on a "privacy shield" to replace the invalidated safe harbour.[1]

The case that drove this decision was filed by Mr. Max Schrems. Schrems is an Austrian graduate student who believed that Facebook sent his data to the U.S. National Security Agency. His case asked that the data protection commissioner of the European Union forbid Facebook from sending his personal data to the U.S. government. Before this landmark case, many governments agreed to a "safe harbor" agreement between their country and Europe, and the United States enjoyed this agreement as part of its overall design for protection of the privacy of EU citizens.[2]

Mark your calendar for January 28. For more than ten years now, about the same time that Google Maps was introduced, twenty-seven countries across the European Union, Canada, and the United States have reserved that day as Data Privacy and Protection Day.

Growing Concern

According to recent surveys, citizens around the globe are concerned about their privacy but are not sure what to do about it. The Rand Corporation conducted a study and found that, across twenty-seven European Union countries, people want more privacy measures in place.[3] Many Europeans will shop around for an Internet service provider (ISP), until they find one with privacy features built into their service.

The top three features Europeans look for are:

1. The ISP can and will actively hide data regarding the user's Internet use.
2. The ISP notifies its users about websites that do not meet their desired level of privacy.
3. The ISP provides practical advice on how to anonymously surf the Web without falling prey to website data collection techniques.

Pew Internet Survey results from 2015 indicated that:[4]

- Ninety-three percent of adults want to be in control of who can access and see their data.
- Ninety percent want to control the information collected about them.
- Eighty-eight percent of U.S. citizens responding to the survey also said they do not want to be observed digitally without prior consent.

Sixty-five percent of those Americans surveyed said there are not enough limits on the data that the government can collect about them from their Internet or phone traffic. Despite the frequency of credit card breaches, Americans evidently trust the credit card companies more than the government to keep their personal details private: 9 percent felt credit card companies could keep their data private and secure but only 6 percent thought that government agencies could.

"[Privacy] *facilitates trust, friendship and intimacy*: qualities that allow us to relate freely to each other. . . ."[5]

Who Is Trying to Look at You Online?

When you are naked (literally or figuratively), it might be a good idea to know who is taking a peek. Certainly there are people who look at others online out of curiosity or to seek titillation. The Ashley Madison hackers dumped information, such as names and e-mails, on the Web for anyone to search and see. The information that data thieves dump on the Internet for the world to search and see can range from the sensational, such as the Ashley Madison case, to the mundane, such as data that are already a part of public record. Everybody from your employer to criminals and spies are also out there, just dying to develop a relationship with your mobile device, computer, your connections, your information, your identity, and your money.

In a study led by Microsoft, twelve countries were surveyed regarding modern technology's impact on society. In eleven of the twelve countries surveyed, respondents said privacy has been negatively impacted by modern technology. The only country where people didn't feel that way was India. Most respondents added that international and local laws have kept pace to protect their personal information.[6]

Anybody can use the information you've left behind to uncover your secrets to hurt you or judge you in ways that may cost you a job, a relationship, your pride, or your reputation. Want to know who is looking? Your school or employer. A competitor. A stalker. An identity thief. Your future wife. The FBI. When it comes to your information online, your motto could be "suspect everyone."

Despite the salacious stories and media attention, criminals may be the least of your worries. The fact is that most people who are looking at your Internet identity already have a relationship with you: a friend, creditor, potential lover, insurer, the tax collector, future or current employer, law enforcement officer, or your government.

To figure out who might be looking at you, think of the primary and secondary relationships in your life, both formal and informal. Casual searches or deep, online investigations tend to be started by:

- those considering a financial relationship with you, such as a car or mortgage lender, your insurance company, your bank, or current employer;

- businesses that want to tailor marketing and advertising, or that want to resell your data to data brokers. These companies may also use your behaviors to design, develop, and refine new products or to detect your human behaviors (for example, your spending patterns);
- people concerned about your reputation, such as your fiancé, potential employer, or school admissions officer;
- those who want to hide behind your face or reputation for their own personal gain;
- retail stores where you shop; and
- people who want to know more about you to bully you, sell to you, or steal from you.

So, who is watching and possibly tracking you? What are their motives? What methods do they use? Let's take a look.

The Workplace: Who Might Be Looking Over Your Shoulder?

Many of us put in long hours and spend time working nights, weekends, and holidays for our employers. Our jobs eat up so much of our personal time that we work during our personal time and take care of personal stuff during working hours: a quick purchase online, checking personal e-mail to see if your home mortgage will be refinanced, or ordering flowers for a sick loved one. Most of us simply do not have another way to juggle it all.

If you've worked for a company for a while, it's likely that your boss is aware that your online usage doesn't interfere with your work and you're not at risk. However, when that new manager comes in or the company takeover brings new policies with it, things could change. It's important that you are aware that your employer has a right to watch your Web habits.

A Quick Lesson in How to Get Fired

You may tend to think of your life in compartments: there's your family life, your life at work, and your social life. Though these compartments overlap at times, you probably think that what you do when you are not at work is your own business. To some extent that's true, but the lines have become blurred in the social networking era.

You carry personal digital devices in your purse or on your belt, and your employer may even provide devices such as a company laptop to get 24/7 digital access to you. More and more, employers are watching what their employees do "off the clock" because they see it as a reflection of their company's brand or an absolute necessity to protect company customers and intellectual property.

Recently, six HSBC bankers in Birmingham, England, created a video of a mock ISIS beheading. The video was posted on Instagram. The bankers in the video were promptly fired. A high school teacher from the Bronx, New York, Chadwin Reynolds, was fired for friending female students on Facebook. After friending the young ladies, he posted comments such as "this is sexy" under their Facebook photo posts.[7]

In the United States, courts have often sided with employers in holding that, at least for an at-will employee, negative work comments online can be grounds for termination. However, the National Labor Relations Board (NLRB) has provided some clarity on this topic. The NLRB ruled that a YouTube video made by construction workers to highlight unsafe work conditions was protected. If your social media posts about work focus on issues with pay or working conditions, your comment may be protected by U.S. labor laws, however, the NLRB has warned that venting about your boss and your boss's annoying habits is most likely not protected. Make sure you refer to your employer's social media policy in your employee handbook before you discuss where you work and say anything about your employer, positive or negative.[8]

All the Wrong Moves

What, exactly, could you do online that could cause you to lose your job? Here's just a partial listing:

- Make racist or sexist comments about your employer or coworkers.
- Mix friends and work: Sending out posts indiscriminately to both friends and coworkers could lead to oversharing or annoying one or both sets of people you know. Consider maintaining separate lists for social networking posts, one personal and one professional. You can even create more than one account on networks such as Twitter. Google+, a Facebook competitor, allows you to create small and large circles of people you know that will make it even easier to keep your posts specific to the most appropriate audience.
- Upload pictures and posts: A picture of you drunk on your Facebook page may seem funny at the time, but don't forget to look at it from your current or future boss's point of view. Posting a comment about how silly and ridiculous your all-day staff meeting was is a really bad idea. A good rule of thumb: If you are thinking of posting something shocking or in bad taste, spend a minute in your boss's or mother's shoes before you do.
- Share secrets: Sharing sensitive or confidential information, if discovered by your employer, is definitely going to get you fired and you might even be prosecuted. Bragging about a special project you

are assigned to is not only bad manners but also a breach of trust. A Microsoft employee bragged about details of an upcoming version of Windows. That kind of indiscretion could prove devastating if hackers were able to design new attacks that would be ready the day the next version of Windows is released, and so the employee was fired. Releasing inside product information from a public company could also be a securities law violation.

- Bad-mouth: Openly complaining on social media sites about how much you hate your job is a bad idea. Do not blog, e-mail, or post on social sites that your coworkers are inane, your boss does not have a clue, or your job is soooo boring. These comments can easily get back to your workplace.

- Play public hooky: Calling in "sick" when you are not is common enough and an integrity issue that we will leave up to your judgment. But if you spend your "sick" day at the beach, don't take date-stamped pictures of you prancing around in your bikini and then post them online. One woman called in sick to her employer, Nationale Suisse, saying she had a headache and could not work at a computer because the light hurt her eyes. However, she did manage to check posts on her Facebook page. They fired her. She contends she did nothing wrong and that she used a smartphone, not her computer, to check her posts.[9]

- Overshare: A woman was dismissed from jury duty after posting that she was conflicted, did not know how to decide the verdict, and that she was taking a poll via Facebook.

- Vent inappropriately and unprofessionally: One teacher lost her job because she had created a blog, wrote anonymously, and posted her negative feelings about her school day, the kids, and more. She never named the students, but someone found the site, alerted her employer, and she was suspended with pay.

How Employers Use Your Information

Sixty-six percent of all employers that responded to a survey from the American Management Association said they monitor their employees' Internet connections when they are using a work computer or an office Internet connection.[10] Today, social media is regularly used to look into your background before an employer even meets you. Many recruiting professionals say that social recruiting is now the primary source of getting to know potential candidates. In a 2014 survey, 93 percent of recruiters used or planned to use information gathered from social media to support their recruiting efforts. According to the same survey, recruiters primarily surfed LinkedIn, Facebook, Twitter, Google+, RSS feeds, and YouTube.[11]

If you log into your employer's network while on a business trip to check your business e-mail and send a résumé to a headhunter while you're at it, your current employers could see that. If they do, you might kiss your current job goodbye.

Employers use your information in several ways, including:

- Keeping tabs on your reputation: employers use the Internet to recruit, do background checks, find out with whom you associate, and more. Once you are hired, you have to stay disciplined about your life online because they will still be checking.
- Hiring, firing, and rejecting: Statistics indicate that recruiters are using new methods to find people to interview and to check out their backgrounds and habits.[12] They look at Twitter, LinkedIn, Facebook, and blogs. Microsoft and Cross-Tab conducted a study and found that human resource professionals have rejected candidates based on what they found online.
- In the United States, 48 percent of employers rejected a candidate because of social media posts. In the United Kingdom, 68 percent of employers rejected a candidate specifically as a result of their social media postings. Twenty-four percent of employers gave warnings to current employees as a result of their social media posts. But there's also a flip side: 15 percent were suspicious of a candidate that had no visible social media presence.[13]

Do you think your employer should notify you about these practices? Think again. Many people assume that employers must disclose their practices if they are monitoring their employees' Internet traffic and e-mail closely. In fact, in the United States only Delaware and Connecticut require employers to notify their employees that they are conducting electronic monitoring and surveillance.[14]

Because of the importance many employers place on Internet and social media reviews in hiring and employment considerations, companies like Social Intelligence Corporation make a business of running these checks on a company's behalf and only reporting certain results back to the human resources department. This service allows a curious employer to run an Internet check without subjecting itself to claims of employment discrimination. A new industry is rising from the business interest in your online persona.

Some information about you can be used to create risk scores that could be used in hiring decisions. The nonprofit organization, World Privacy Forum released findings that various third-party firms and companies have created "secret" consumer scores that rank you on everything from the likelihood you will keep your job to how likely you are to commit fraud. One company sells a score that uses employment and unemployment data, as well as economic trends and forecasts to predict the probability that you will lose your job, and, as a result, not be able to pay your bills.[15]

Your employer may or may not know the details behind your risky score but the Internet knows all and never forgets. In the interest of self-protection and not knowingly violating existing laws, many employers prefer to know which candidates would be costly to employ, even if they don't know the specifics. Although they do not need to find high-tech solutions to do so, using external scoring leaves employers less vulnerable to questions about their hiring practices, particularly if they don't appear to be actively looking for medical issues.[16]

Clients See You, Too

Do you wonder why your biggest client took his or her business and walked away last year? Sometimes a client might not hire you in the first place, and you're not really sure why. Or you could have a long and profitable relationship, and one slip or careless post could leave you without a client.

FROM THE HEADLINES

According to a story from Eric J. Sinrod on CNET from May 30, 2007, David Mullins was fired from his job as facilities engineering technician for the National Oceanic and Atmospheric Administration's Weather Forecast Office in Indianapolis. He filed a claim against his former employer stating that his supervisor was improperly prejudiced against him when she "Google searched" his name and discovered that he had been terminated from his previous job with the U.S. Air Force. Mr. Mullins's supervisor ran an Internet search and found this information: "In 1996, the Department of the Air Force removed [Mr. Mullins] from a civil service position and that in 1997, the Smithsonian Institution told [Mr. Mullins] to 'look for a new job.'"

Mr. Mullins claimed that such a search into his past, and subsequent termination of his federal employment based on the search, was improper under U.S. civil service rules. The board hearing his case, and the Appeals Court in review, left this question open, finding that he was properly let go for other reasons. Yet his supervisor clearly looked up Mr. Mullins's history online and her findings could easily have played a role in his firing.

Look at the situation that comedian Gilbert Gottfried, the iconic voice of the Aflac duck, encountered. He posted offensive and inappropriate tweets regarding the Japanese earthquake and tsunami crisis. After those tweets gained the attention of the press, Aflac terminated their relationship with him. Aflac's senior vice president and chief marketing officer, Michael Zuna, said in a company statement, "Gilbert's recent comments about the crisis in Japan were lacking in humor and certainly do not represent the thoughts and feelings of anyone at Aflac. Aflac Japan—and, by extension, Japan itself—is part of the Aflac family, and there is no place for anything but compassion and concern during these difficult times."

A media company lost their client, Chrysler, over a tweet. An employee of the media company, called New Media Strategies, posted the following, "I find it ironic that Detroit is known as the Motor City and yet no one here knows how to (expletive removed) drive." That employee cost New Media Strategies a client.[17]

Where You Shop: Who Might Be Looking Over Your Shoulder?

We expect to get help when shopping, but we may not realize that we are being watched by those who provide that help. Surveillance of shoppers is nothing new. Many retail outlets observe shoppers to see who might need help, identify the times of day that require more or less staff, or offer you just-in-time special offers on your phone when you connect to WiFi. But did you know that many of the stores that you visit are photographing; they may or may not keep those photos on file.

Several retailers, in an effort to fight crime, are using facial recognition technology. Walmart has been testing technology that may help them identify shoplifters in their stores.[18]

If a face matches a database of a shoplifter, the system alerts security. The problem with this is that facial recognition technology is not perfect, and some of us have doppelgangers. It's also possible that in the future, at the payment terminal, stores may scan your face and match it up to social media posts and other sources to validate your identity. Though this may sound like a great way to protect your identity, what happens if that database, storing your facial recognition photos and tied to your credit card, is hacked?

What should the expected privacy policy be in these cases? No customer is asked to review a store's privacy policy and click on "ok" before they enter the store. Still, what happens in these brick-and-mortar encounters can draw information about you from the Internet, and post information about your shopping activities for others to view. If you think you're not "naked online" because you're careful or "off the grid," think again.

People in Your Personal Life

Quite often it is somebody you know in your personal life who exposes or uses your personal information in ways that can harm you. Here's the rundown of relationships you might want to review with this in mind.

Parents Are Watching

Social networking sites, e-mails, and texting are a great way to stay connected to family members. Some parents keep up with what their kids are doing online, which is the responsible thing to do. Sometimes kids see this as intrusive surveillance, so be sure you set ground rules early on, the younger the better. Try to make your kids feel that you want to share their experience of learning about the Internet just as you help them with homework, and not to invade their privacy. One mom regularly logged into her sixteen-year-old son's profile on Facebook to make sure he was following the home rules. One day while checking his page, she did not like his posts, so she removed them and changed his password.

The son got upset and filed a complaint with prosecutors, who agreed to hear his case against his mom under the state's harassment law.

Kids Watch, Too

Don't forget that your kids are watching you and other family members online, too. They are searching social networking sites to see what their parents post. Some kids are using social networking sites to find and reconnect with a parent they may have been alienated from as a result of divorce or other family changes. Some kids are searching their parents' posts to justify what they post, so if you post anything online, be sure you are the role model for the behavior you espouse.

If your child tends to be disrespectful to any family member, make sure you monitor their online communications with those family members. The elderly, especially grandparents, are often considered weaker and may be the target of the family bully. Kids may use social networking and texting to abuse their grandparents or to bully grandparents into giving them money or covering for the kids when they get in trouble.

Best Friends Forever (BFF): Friends Online

Your friends could inadvertently expose your information in their own online postings or through online activities such as social networking or genealogy. Also, relationships change, and a friend today could turn on you as a cyberbully tomorrow.

But it may not just be your current set of friends exposing you. It used to be that when you grew up and moved away from home, you had the opportunity to make new friends and only keep up with old friends if you wanted to. Now, on the Internet, you could be reconnecting with people you want to and with those who you would rather forget. There are TV and radio ads that encourage you to find old friends or former loves online. There are, in fact, many websites dedicated to helping people find other people. Just as you might have brushed off such a person who called you out of the blue in the past, you have to learn how to manage who you do and do not want to associate with online.

Prospective Mates and Spouses Are Watching

Unearthing information that used to take families lots of money, several months, and a professional investigator might be just a few minutes and mouse clicks away today. You might be the one to perform such research or be the subject of research by others. Worried that your mom's new boyfriend is really after your deceased dad's bank account? Start surfing the Net to learn more. Is Mr. Wonderful starry-eyed in love with you, but some of your facts and stories don't quite add up? You might be in for a nasty surprise when he breaks the engagement.

Here's one example of what a spouse might find out online. A woman from Ohio suspected something might be wrong with her relationship with her husband. She went to Facebook, typed away, and found her husband appearing handsome and happy on his wedding day. He looked fabulous, and he had posted roughly two hundred photos online. Sounds good so far; the only problem was that the bride was somebody else.[19]

Criminals

Criminals go where the action is, and there is currently a lot of action online. Criminals used to case a bank or cruise around looking for victims, but today they find it far more efficient and effective to conduct some or all of their unsavory research online. They have many methods of finding you, watching you, and tracking you online. They surf social networking sites, chat rooms, and gaming sites. They also have mastered techniques for looking like a legitimate company, friend, or associate online. They use information they have discovered about you to implement tactics that gain your trust, and they try to trick you into clicking on a link or giving up information you wouldn't hand over to a stranger.

Cybercrime is a global issue that is on the rise. It will continue to be attractive to thieves until strict laws are put into place and there is transnational cooperation among governments, international law enforcement agencies, and international court systems. The free and open Internet that has been the fuel behind growth, innovation, and removal of barriers could become severely damaged by rampant unresolved crime.

According to Dr. Sally Leivesley, former Home Office scientific adviser in the United Kingdom, "What we need is an international agreement between all the [United Nations] countries. You have to have a respected highly intelligent body, paid for by the UN to monitor and advise a specific group within the UN when breaches have occurred."*

* James Fielding, "EXCLUSIVE: Cyber Hackers Are GREATER Threat to UK Security Than Nuclear Weapons," Sunday Express, October 25, 2015. http://www .express.co.uk/news/uk/614417/cybercrime-UK-talktalk-hack-security-computer -systems-online-safe.

The types of crimes listed in this book are just a sample of what's happening to innocent victims around the globe.

Social Engineers

Did you know that many security breaches have been attributed to human error?[20] Almost 95 percent of breaches recently in the headlines are attributable to human error and 78 percent of those breaches involved tricking the user rather than using technology. If the bad guys can trick people who are trained to be on high alert, what does that mean for the rest of us?[21]

Social engineering is a popular tool of the trade for both offline and Internet criminals. This tactic involves gathering information about you to take advantage of you. Criminals may purchase information from other criminals following a data breach or they may go to free, online records, databases, and social media to acquire the information.

What do they look for? Information they can use such as your hometown, favorite hobbies, high school mascot, and family member names. Then, they use that information to gain your confidence and trust. This happens on social networking sites, dating sites, in e-mail messages, text messages, gaming sites, and phony websites that download malware to your computer with every click.

Most people leave a treasure trove of clues for criminals who are planning to commit social engineering crimes or physical robberies. A ring of robbers in New Hampshire used Facebook to determine what homes would be vacant and how to plan their robberies. If you're leaving enough information around online, social engineers might use it to deceive you, stalk you, or rob your home.

Clicking Your Way to Trouble

Criminals are becoming increasingly sophisticated at taking information from your Internet identity to convince you to click on links, open attachments, or fill out forms that give them the information they need to take over your computer or your online identity.

E-mail Alert

When you get those annoying e-mails about drugs you do not want to buy or e-mails from vendors you don't do business with, it's possible that you are being spammed. By clicking a link or attachment in the spam e-mail, you might just be downloading malware. Malware is malicious software that is downloaded to your computer or other device. Criminals use malware for many purposes, but the basic intent is to follow you around online so they can collect information about you or to gain access to information or damage data stored on your computer.

Some criminals create programs that "wake up" to steal your account ID and password when you log into an online banking site. Some criminals create programs that use your own address book to send out spam to your friends using your name, hoping to pull in more victims.

You may accidentally click on what you think looks like a virus software update, also known as *shareware*. This invites criminals right into your computer. Shareware criminal rings are big business in the Internet world. The FBI, in coordination with many other countries, was able to bust up two cybergangs using scareware to steal over $74 million.

Pay the Ransom

A popular scheme today is the use of ransomware, which cybercriminals employ to infect your computer or the network of a company. They silently track your files to determine which data might be most valuable to you. When the cybercriminals strike, they encrypt important files and threaten you: If you don't pay their ransom, they won't unlock your files and in some cases, they threaten to destroy the files.

Use of ransomware was up more than 100 percent in 2014 from 2013. Many security and law enforcement experts believe this number is too low because victims often pay the ransom and don't report it. Security software firm Symantec estimates that at least $5 million was paid by people and companies responding to ransom e-mails each year and payment of ransoms by victims is now on the rise.[22]

The FBI has estimated that roughly $27 million in ransom demands were paid just for the victims infected with the CryptoLocker virus during the months of September 2013 through May 2014.[23]

Thankfully, security firms eventually find a fix for each instance of ransomware, but the scheme is so lucrative that cybercriminals develop new approaches designed to get around the defenses of security products such as antimalware and antivirus software.

Smishing

While phishing e-mails that pretend to be from a source such as your bank are commonplace, smishing is moving up on the list of Internet crimes that involve text messaging. *Smishing* is a phishing message sent via SMS messaging, which is typically available on most mobile phones and tablet devices. One click on that smishing message could infect your phone and allow criminals to peek at your address book and surfing activity.

A 2014 report by EMC indicated that all phishing scams are costly. Roughly $5.9 billion was spent recovering from 500,000 attacks.[24]

There are seven steps to getting your account back from phishing cybercriminals or preventing cybercriminals from owning it in the first place:

1. *Service providers can help.* If you're locked out of your account, you should try to regain access from the service provider. Go to the help area of your provider. There are some quick reference links for Gmail, Hotmail, AOL, and Yahoo!
2. *Change your password.* If somebody does grab your account but you can still log onto it, immediately change your password. See a demo for how to choose a strong password at http://www.wbtv.com/story/27286716/

IT'S A GLOBAL THING

Websites that house malware and viruses originate from countries around the globe. According to researchers at G DATA Security Labs, the majority of malicious and fraudulent websites are hosted on servers in the United States (a whopping 43.3 percent), followed by 9.5 percent in China, and 8.2 percent in France.*

---

* Sara Peters, "Report: One-Quarter of Malicious Sites Healthcare-Related," *Dark Reading,* October 22, 2015. http://www.darkreading.com/analytics/report-one -quarter-of-malicious-sites-healthcare-related-/d/d-id/1322795.

facebook-searches-for-stolen-passwords, or try out the Haystack tool at https://www.grc.com/haystack.htm.

3. *Check your settings.* Once you get back into your account, check all your settings. The hackers may have changed settings such as "forward all e-mails to account ABC@differentmail.com."

4. *Check any auto login from e-mail to other apps or accounts.* If you auto login for other apps or accounts, you may want to change the passwords for those apps and check their settings as well.

5. *Tell your friends and family.* Tell everyone your account was hacked and not to respond to e-mails from you, click on links, or open attachments.

6. *Use two-factor authentication.* Before the worst happens, try turning on two-factor authentication. Two-factor authentication is usually tied to your cell phone and many Internet companies will send a short code to your cell phone when they see a login that doesn't match your usual patterns (e.g., different location or a different device).

7. *Run antimalware and antivirus software.* Make sure you don't have malware logging your keystrokes; if you do, the cybercriminals will be back! Microsoft has a free malware removal tool and Sophos has one that works on both Macs and Windows-based computers.

If you need to get your e-mail account back from cybercriminals, go directly to your e-mail provider for more information.

Criminal Techniques du Jour

The profile of the criminals behind spam, scams, and Internet identity theft is hard to pin down. Law enforcement officials have uncovered sophisticated networks, but they have also found criminals who acted alone. But no matter what the criminal profile, these creeps have a variety of techniques you should know about to protect yourself.

Nigerian Scams

Advance fee scams, also referred to as "419 Internet scams," are often tied to Nigeria. The 419 is a reference to Nigeria's criminal code for fraud. These are essentially phishing scams that usually involve money. The criminal has your e-mail address and sends you a message begging for your help and assistance. According to Krebs on Security, there are various auction sites that help these criminals get your e-mail address so they can send you their scam e-mails. 419eater.com is a group of people who have come together to track these 419 scams to spread awareness and alerts to the general public.

Spyware

Spyware finds its way onto your digital device by hiding behind free apps or ringtones or games, pop-up ads, or through scareware messages that tell you your computer is being repaired while actually installing software to track and report your every move back to the criminals. Spyware is sometimes used to get information for marketing to you, but criminals could also be using it for more sinister activities. Internet criminals find spyware an attractive method for collecting information such as e-mail account logins, websites you like to surf, and instant message contents.

Even legitimate looking apps can fool you and the online store you buy them from. In September 2015, the popular Apple App Store had to remove several apps that had become infected with malware. The malware hidden in the apps was designed to trick the user into giving up passwords to their Apple iCloud accounts. In October 2015, Apple found that another set of apps had to be removed. In this case, Apple found that roughly 250 apps were collecting the personal information of anyone who downloaded them. Many of the apps were, in some cases, downloaded as many as one million times.[25]

The Google Play store has had its share of scamsters, too. In early 2015 there were a few Android apps that behaved normally when first downloaded. After thirty days, however, the apps would begin spamming users with messages that their computers were out of memory and that they had a security issue. The evil app would then reroute them to other malware-laden web pages and apps that could supposedly help them resolve the issue. The motive for this scam was to collect information about the users.[26]

Poisoned Search Results

Criminals know that people trust and use search engines such as Yahoo!, Bing, and Google. Google reported that roughly 1 percent of its search results might contain poisoned links, which are links that take you to malicious sites. Google does its best to manage and filter those criminal links out, but it's a good idea to think before you click on those search engine results or a link within the results. Sponsored ads that appear at the top of search results are often the most likely to contain such dangers.

Clickjacking

Clickjacking is a variation on using social engineering to trick you into clicking on a link. The link appears to be legitimate because it seems to lead to a business you have visited or an item that should interest you based on what the criminal has learned about you. In reality, the criminal has hijacked the page so that your click activates the code they want to execute. You probably won't know what has happened, and you might keep clicking away on a site that you think is Facebook without realizing that you have been clickjacked. A typical tactic is having the clickjacked link suddenly sent out to all your contacts so that it becomes viral, affecting expanding groups of people in a geometric pattern. Criminals like to use clickjacking to steal your personal information, but sometimes they hijack clicks to earn money on surveys that pay by the click.

Social Media Posts Could Be Your Way Intro Trouble

Companies now encourage us to tweet, Facebook, Google+, and Instagram them when we have a good or bad service experience. The latest cybercrime scam waits for people to post on social media that they are upset about a customer service issue. Scamsters then send the individual a customer service message that looks like it is from the legitimate company. The consumer responds thinking their issue is going to be handled and accidentally gives away critical information about them.

Sexual Predators

A Pew Internet study found that one in five sixteen year olds has received a sexually explicit text, or a sexting message. That number jumps to 30 percent when teens reach age seventeen. An *Archives of Pediatric & Adolescent Medicine* study states that 25 percent of teens have sent a sext.[27]

According to the National Center for Missing and Exploited Children (NCMEC), in 2014 their CyberTipline received 1.1 million reports, most of which related to child sexual abuse images, online enticement, including "sextortion," and child sex trafficking.[28]

Grown women are targets of sexual predators, too, with the predators' primary outreach on social networking sites and online dating spaces. Congress is so concerned about online safety for both men and women on dating services that it has introduced draft legislation requiring dating sites to tell their users whether they require background checks. Parry Aftab, an Internet privacy lawyer, warns, "No central authority or group is counting how many sex crimes are Internet-related."

Online sexual predators have found a twenty-four-hour playground to target victims. They're watching in chat rooms, they barge in on instant chat sessions that do not have privacy settings turned on, they frequent social networking sites, and they play games in Internet gaming forums.

So, who are these people? The profile of sexual predators may surprise you. The Center for Internet Addiction Recovery reviewed twenty-two cases of alleged sex offenders and found that Internet sexual predators typically have an addictive disorder, and they use online sexual exploits as a way to avoid facing issues in their personal lives. According to the review, sexual predators are typically male and range in age from eighteen to fifty-five, but they can be as young as thirteen. Some of the predators are married. Law enforcement says that Internet sexual predators know what they're looking for and how to connect based on their personal preferences. They especially like to find victims who post notes of frustration, loneliness, or sadness because they can more easily befriend them and earn their trust over time.

In one real-life example of sexual predation, a forty-one-year-old male tried to befriend a fourteen-year-old girl, who he found on Facebook. He used Facebook to message and text her. He encouraged the girl to sneak away to meet with him and finally convinced her to come. Only in this case, thankfully, the girl was actually a police officer. Law enforcement was given a tip, and they took over her Facebook account and communicated with the man using the fourteen-year-old's online persona. He met his "girl" and went off to jail. Unfortunately, not all of these criminal actions end this way.

The issue of online sexual predators has gained a lot of attention. For example, actor David Schwimmer from the NBC comedy *Friends* produced a film called *Trust*, which addresses the issue of Internet sexual predators. He felt he needed to do something to open up the dialogue between parents and kids about the dangers of being tracked and approached online. You can watch a promo for the film at www.youtube.com/watch?v=6qDwNCzlidI.

Revenge Porn

Revenge porn is legally defined by many countries as the sharing of personal, private, sexually explicit photos or videos of another person without their consent. Often this is done with the goal of either extorting the victim or causing the victim personal distress.[29]

The crime is increasing, motivating countries around the globe to enact revenge porn laws. England and Wales passed the Criminal Justice and Courts Bill that includes an amendment to address revenge porn. If you are convicted of trafficking in revenge porn in these countries, you could face up to two years in prison.[30] Several U.S. states have revenge porn prohibition statutes as well.

Many jurisdiction's laws have not kept up, and there is little hope for victims of this horrible crime. The good news is that social media networks are standing up for revenge porn victims. In February 2015 Reddit, a digital social media bulletin board, said it would ban all sexually explicit content posted without the consent of those in the pictures. Twitter enacted something similar in March 2015. Twitter now immediately removes any "link to a photograph, video, or digital image of you in a state of nudity or engaged in any act of sexual conduct" that has been posted without consent. In June 2015, not long after the Reddit and Twitter changes were put into effect, Google stated it would delete links to revenge porn on request. In July 2015, Microsoft created an online form for victims to fill out to request removal of revenge porn.[31]

VICTIM PROFILE

Dahlia*, aged twenty-one, was in her last semester attending New York University on a full scholarship when she received the first text message. It was from a number she didn't recognize and read, "ur ad iz so hottt. watz ur rate?" The sender then went into graphic detail complimenting Dahlia's body, accurately describing her hip tattoo, and describing the things he wanted to do to those hips. Five minutes later, her phone vibrated with another message from a different number, this one containing a picture of somebody's penis. The texts—and then calls—didn't stop coming and were all from different callers. Some seemed to know her name and the address of her dorm. Finally, one sender took mercy on her and sent a link to the ad everybody was referencing. To Dahlia's horror, there on Backpage.com were naked photos of her advertising sex: "Shy College Girl but Ready, Willing, and Blow-ur-soxxx-off Kinky—Outcall Only—21 (NY)." The ad contained her phone number, address, the name of the gym where she worked part-time, and school. Dahlia flagged the ad, and it was removed by the site administrators, but almost immediately a new one replaced it. Then the naked images began popping up other places—on a Tumblr page. Imposter Facebook and Twitter accounts showcasing the nude images were created in her name and then friend requests and follows were sent to Dahlia's friends, relatives, classmates, and even her professors.

Over the course of the week, Dahlia's search engine results changed. Her LinkedIn page was still up top, but the next ninety-three results were populated not by articles about her semester working at an NGO in Mali or the award she'd won for leading the Model United Nations. Rather, the hits led to links to her nude pictures on Internet sites she never knew existed: Xhammster, Pornhub, Xtube, and Redtube.

Dahlia tried to remove the images, but it felt futile—like a game of whack-a-mole. The images were spreading faster than she could remove them, and some websites refused to take them down without her sending them letters. She got some help sending those letters, but then the letters with her name and the link were published on a website managed by Harvard and the pictures still existed on the Wayback Machine, an Internet archive. Some sites refused to remove the pictures unless she proved she owned the copyright. Unfortunately, some of the pictures were taken by her ex-boyfriend. Technically she couldn't claim copyright ownership of those, even though it was her naked body depicted.

Dahlia tried keeping it in perspective. Unlike the poor people she'd helped during her semester abroad in Mali, at least she had clean water, electricity,

(*continued*)

39

a bed. However, life was different now. Even basic things like grabbing coffee at the Starbucks next door or going to her favorite kickboxing class felt horrifying and exposing. By visiting the porn sites where her pictures were posted, she could see that half a million strangers had seen her genitals, not to mention those known to her who had received direct contacts over social media. Meanwhile, her family and friends were less than supportive. She'd stopped counting the number of times somebody had said, "That's why *I* don't take sexy pictures" or "You know, you do share some of the responsibility."

Most difficult of all was that Dahlia did not know who did this. Only one other person had ever even seen the pictures, Thor. In fact, he'd taken some of them. Their relationship ended on bad terms, but that was months ago and when she reached out to him, he assured her he wasn't responsible. Who could it be, though? Who could hate her *this much*? Was she hacked like the celebrities during the summer of 2014? There was the time she took her computer to the Apple Genius counter after the Mountain Dew mishap. That guy was sketchy. What about that creepy guy she met on OKCupid who told her he would destroy her life when she turned him down for a date. And she never did find her old iPad after her roommate had that party. Oh, and her phone was lost in an Uber for twenty-four hours before she began password protecting it. OMG, had she remembered to log out of her iCloud account at the school library? At the computers at work?

Meanwhile, life was getting worse. More strangers were calling and harassing her. All the jobs at NGOs and nonprofits where she'd interviewed before this began, stopped pursuing her, and nobody was inviting her on interviews. Dahlia went on a couple first dates, but as soon as the guy learned her last name and Googled her, there were no second dates. Or if there were, it was because he thought she was a porn star. And truthfully, that's how she felt—like an indentured porn star, forced into becoming the sexual entertainment of others. The counseling center at school was supportive, but they seemed most interested in making sure it was not a school-related event that would trigger Title IX. Somebody suggested she tell the cops, but when she went to the precinct, the officer brushed her off, telling her to just stop being on the computer all the time and informing her that there's nothing illegal in her state about sharing consensually created naked pictures of an adult. Plus the offender could be on the other side of the world, they told her, adding that law enforcement is busy enough fighting *actual* danger.

---

* Names have been changed to protect the victim.

Interview with an Expert

Revenge-porn lawyer Carrie Goldberg, founding attorney, C. A. Goldberg, PLLC

Q: Can you describe the legal work you do for victims of Internet privacy invasions and revenge porn?

A: I represent individuals under attack. I sue rapists, get justice for victims of revenge porn and sextortion, protect public figures from psychotic superfans. I'm the go-to person if a person's personal information, sex videos, or photos are going viral on the Internet. My most serious cases are victims who are underage. Often the defendant is the school.

Q: What kinds of problems do you solve that are unique to this practice area?

A: Confirming the opposing party is a major first issue—because many on-line attacks are anonymous. We do intelligence and determine threat risks. Is the offender a present danger? Additionally, is the victim a danger to herself? Many victims do not know how to cope with nonconsensual Internet exposure. We remove content from the Internet, too. And of course, we sue when needed. Because these are typically highly private matters, we make pre-filing motions to proceed under a pseudonym and for the file to be sealed.

Q: What are common scenarios you see relating to Internet privacy?

A: They run the gamut. Jilted ex-boyfriends who feel the need to "settle the score" by sending naked pictures of their ex to everybody in her social circle. We see ex-mistresses who are threatening to spread images if the sugar daddy refuses to pay up, celebrities who have superfans stalking them, teens whose nude images have circulated around the school. I also have clients who were extorted by strangers who befriended them on the Internet. Sometimes offenders purchase ads on Craigslist and Backpage impersonating the victim soliciting sex and information about how to contact the victim. In fact, the majority of cases involve the publication of truthful facts (e.g., name, address, school, date of birth, phone number, e-mail, employer, etc.) about our client alongside the intimate images which make victims prone to harassment from the revenge porn consumers. In many cases, the image was not created with consent—the victim was photographed or filmed without her knowledge or she was passed out or her webcam was hacked and used as a portal to film her. Worse still are the scenarios in which the recorded sexual act was nonconsensual. The saddest cases are those in which a rape video is going viral around a middle or high school.

Q: What sparked your interest in this area of law?

A: Nowadays, if a person is under attack, almost always there is an online component to that attack. My knowledge base is unique—16 years of working with traumatized people, a litigation background, experience lawyering in the fields of domestic violence, Internet law, intellectual property, knowledge of criminal and first amendment law, etc. Most importantly, I can personally relate to what it means to be under attack. I started this firm after I was the target of a malevolent actor and was unable to find an attorney who could help me. My only objective is to be for others the kind of lawyer I needed during my turmoil.

Q: What tools do you use to help victims of revenge porn?

A: My plan of attack varies based on the priorities of the victim, the information we have about the offender, and the nature of the attack. There is no proscribed way to handle these situations and instead we must carefully consider all factors at play. There are generally six categories of tools I use:

1. ISP advocacy
2. copyright
3. criminal law
4. family offense proceedings
5. civil court
6. school disciplinary proceedings and Title IX

Each of these tools is imperfect, but collectively they provide victims of online harassment with a wide range of options.

Q: Can you briefly describe each of those six tools?

A: Yes.

- Internet Service Provider Advocacy: Offenders frequently use mainstream social media companies to disseminate images, create imposter accounts in the victim's name, and for delivering threats to victims. In 2015 a number of our biggest social media companies (i.e., Twitter, Facebook, Tumblr, Instagram, Reddit, etc.) established bans on revenge porn. Using forms on these companies' websites, victims' nude pictures can be removed within hours. I have contacts at these companies who I speak to when the form submissions do not result in removal. Search engines (i.e., Google, Bing) also now have forms victims can use so that revenge porn content is removed from a victim's search engine results. I also deal with the ISPs to report abusive accounts. When the revenge porn is published on a website the offender has created, I report it to the hosting company.

- Copyright: Individuals automatically own the copyright images they took themselves. So, if the nude image is a "selfie," any third party who publishes it without the copyright owner's consent is infringing upon that copyright. A victim need not actually register the copyright in order to own it. If a website is notified that they published infringing material, they can be sued pursuant to the Digital Millennium Copyright Act ("DMCA") if they fail to swiftly remove content. Even if the image is not a selfie, I can help negotiate for the copyright to be transferred to the victim so that she can benefit from the DMCA laws.

- Criminal Laws: More than half of all states have laws criminalizing revenge porn. Coercion and extortion laws are also relevant in the frequent cases where the offender threatens to distribute the images prior to actually doing so. So, sometimes, the swiftest and cheapest way for a victim to seek protection is through local law enforcement. I help victims prepare their evidence and advocate for an investigation to be opened into these cases. It can be an upward battle convincing under-resourced law enforcement departments to prioritize these cases.

- Family Offense Proceedings: So often the offender is a jilted ex who is aggressing and stalking the victim in a myriad of ways online and offline. I can commence a case in family court and the victim can leave court that very day with a Temporary Order of Protection. Once served, if the offender aggresses against her, he can be arrested.

- Civil Court: Victims can sue the offender for money or for an injunction to stop him from aggressing. While our criminal laws aim to punish the offender, our civil laws focus on restoring the victim financially. Depending on the state, we can sue for intentional infliction of emotional distress, privacy invasions, and now a growing number of states have specific revenge porn laws. Through civil court, we can also subpoena online service providers to de-anonymize offenders. Through civil court we can also seek an injunction to stop the abuse.

- School Disciplinary Proceedings: For my younger victims of revenge porn, the dissemination of images happens at school. It is critical that the school conduct a prompt and thorough investigation immediately to stop the spread of the image and punish the wrongdoers. Schools must also refer the victim to a Title IX coordinator and make sure the victim is supported. If they botch their responsibilities, creating a hostile environment for the victim, the school risks becoming the target of a Title IX discrimination complaint with the Department of Education Office of Civil Rights or the defendant in a federal lawsuit alleging a Title IX violation.

Q: If you had one piece of advice for the public, what would it be?

A: Nine out of ten of my cases begin with "I met him online." So, be careful with online dating. It's the best thing that has ever happened to predators.

Q: What is a common misperception that people have about victims of revenge porn?

A: That the victim is always female. Although the majority of victims are female (which is why I tend to use female pronouns), plenty of men also suffer from their nudes going viral.

Q: If you had one piece of advice for non-victims?

A: Do not blame the victim. She (or he) is already unhappy and does not need additional judgment. Just as we do not shame burglary victims for living in a house, there is no point blaming victims for taking the picture. Nobody is 100 percent protected from being the victim of revenge porn. Many of my clients never took a nude picture in their life. Rather, they were filmed without their permission, their picture was Photoshopped onto a porn star's body, or their bikini was Photoshopped off of an innocuous beach photo on their Facebook wall.

Q: What are the biggest challenges you face as a practitioner?

A: Anonymous offenders.

Q: How do you address this challenge?

A: I first exhaust all open source information gathering as I can, based on what I know about the offender—usernames, IP information, and whatever personal information he has revealed to my client. If that does not yield results, I retain high level cyber intelligence professionals to help. The next step, if we are still at a loss, we can initiate a civil proceeding against a John Doe defendant. Depending on the state, pre-litigation discovery may be an option too. Once we have an index number, we can subpoena online service providers to obtain meta-data, IP addresses, and other identifying information. Once we have the identity, if we still wish to proceed with the case, we can amend the complaint to substitute the name of the offender.

Q: What's the first thing someone should do if she finds a private photo of herself online?

A: If somebody finds a private photo of herself online, swift action is essential. Immediately take a screen shot. If there's info about when it was posted, who posted it, and how many views it's gotten, take a screen shot of those things, too along with any comments posted. I use a third-party remote system (Page Vault) to document pages so that they are easily authenticated in court. Do not erase any incoming messages from the offender. Back-up and store this content.

Next, contact the website that posted it. Inform them that it depicts you and that it was posted without your consent. Demand that it be removed immediately and that they provide you with the following information:

- Identity of the person who posted image(s);
- IP address of the person who posted image(s);
- Date when image(s) were submitted;
- Date when it was posted (if not obvious);
- Number of unique views the image(s) had received. Save the e-mail and re-send it if the material is not removed.

Search elsewhere online to see if images were posted on other websites. Visit the site regularly to see if it has been removed. If it's not removed within 24 hours, contact an attorney. The nonprofit End Revenge Porn has vetted attorneys who have volunteered to assist victims. If you know who posted it, keep track of all other hostile contact from that person. Contact the police if you feel threatened. Keep notes about your emotional and physical responses (i.e., trouble sleeping, nausea, inability to eat, vomiting). Do not block the offender's phone number. If you have a claim of harassment, the more incoming communications, the better. Under no circumstances, should the victim engage in any direct negotiations or back-and-forth with the offender. Victims should also batten-down the hatches on their online security and privacy—hide Facebook friends lists and require notification if somebody wants to write on your wall. Create complex passphrases and 2-step notification on all accounts.

Q: What resources are available to victims?

A:

The Cyber Civil Rights Initiative (www.endrevengeporn.org) is a nonprofit which has informational resources for victims, links to lawyers, words of support, links to forms for removal on social media, and also provides direct services through a 24 hour crisis helpline (1-844-878-CCRI).

Woman Against Revenge Porn (www.womenagainstrevengeporn.com) is run by a victim and is devoted to helping revenge porn victims through photo removal and other tips and attorney links.

Without My Consent (www.withoutmyconsent.org) is a nonprofit that empowers victims of online privacy invasions to fight against online harassment. It is a great resource for practitioners as well, with terrific documentation of the law.

C. A. Goldberg, PLLC (www.cagoldberglaw.com). My website has information about options available to victims of Internet harassment, revenge porn,
sextortion, and other kinds of invasions of privacy and sexual consent. It has
a map with links to civil and criminal revenge porn laws, a page with links to
social media take-down pages, and a blog.

The Cyber Civil Rights Legal Project (www.cyberrightsproject.com) was founded
by international law firm K&L Gates and provides pro bono legal representation to victims of revenge porn.

New York Law School is the home of the Cyberharassment Clinic which is part of
the Tyler Clementi Institute for Safety and provides pro bono representation
to victims of cyber harassment.

Authorities

Today we have cameras watching us just about every place we go. The
United Kingdom has roughly 1.85 million surveillance cameras, and the
UK's Association of Chief Police Officers estimates that the average person
in Great Britain is caught on surveillance cameras an average of seventy
times per day.

That level of surveillance is not unique to the United Kingdom. For example, the Highland Village police department in Texas uses public safety cameras
to scan license plates to find stolen cars or criminals, and they decided to keep
those images in a database. They have pictures of everyone's license plate on
camera along with each vehicle's location and the date and time that the picture
was taken. They use this database as a primary source for police investigations.
But could that information invade your privacy?

Law enforcement, in an effort to expand their ability to promote neighborhood awareness, is making use of social networking sites and e-mail with great
success. Your neighborhood watch has gone global and digital. If you have an
embarrassing run-in with the law, your misfortune might be digitized for all
your neighbors to see online.

One police department in Ohio is using Facebook to ask citizens for their
help in fighting and solving crimes. They post clues, information, photos, and
videos. If you happen to be in the wrong place at the wrong time, might you,
even if you're completely innocent of any wrongdoing, come under suspicion?
It's important to note that many of the companies that you do business with each
and every day receive law enforcement and government requests for data. Those
requests only seem to increase with time. Facebook, for example, stated on their
blogpost that global government requests for data increased by 18 percent in
the first half of 2015 compared to the first six months of 2014, bringing the
total number of global requests for the first half of 2015 to more than forty
thousand.[32]

Government

Some government agencies get to know all about you. Though you may live in what you consider a free and open society, you might be surprised to learn how even the most liberal of governments treats its citizens' information.

An Issue around the Globe

As of the writing of this book, several countries are debating important rules that will have an impact on your right to privacy and your ability to protect your Internet identity for years, and maybe for your lifetime.

Russia has demanded that Twitter stop storing tweets and profile information provided by Russian citizens in any country outside of Russia.

The situation is so confusing that Microsoft asked their company's chief legal officer, Brad Smith, to blog about privacy following the overturn of the safe harbour arrangement between the European Union and the United States. The EU data safe harbour was part of a regime that allowed the United States to transfer and store EU citizen data on U.S. servers. Smith asked the United States in his blog post to agree that "it will only demand access to personal information that is stored in the United States and belongs to an EU national in a manner that conforms with EU law, and vice versa."[33]

Secrets Out in the Open

Do you support a political candidate? Your employer may have a strong political bias that is the opposite of yours. Many of us have been raised to believe that it is taboo to talk partisan politics in social or work situations. However, if you have donated money to a political campaign, you should be aware that information about your contribution is available for everyone to see.

You might think it would take a covert computer geek to look up this type of information, but the reality is that your neighbor, potential date, and boss can all go online and see what causes you contribute to within minutes. One of the more popular sites to check out this type of information is www.OpenSecrets .org. You can also look up your favorite or not so favorite politician to see who their major contributors are. So much for keeping your political affiliations a secret.

Here's a story about who is watching you, where the virtual world meets the physical world in a case of old-fashioned tracking.

According to a story covered in *Wired* magazine, twenty-year-old Mission College student Yasir Afifi was getting an oil change when the mechanic called him over to look at something odd on his car. The FBI had warned Afifi a few months earlier that, because of an anonymous tip, he might be watched as a national security threat. Sometime later, Afifi's mechanic pointed out the surveillance equipment under his car.[34]

Afifi found two devices near the car exhaust attached by magnets. He removed these devices and photographed them. Then Afifi posted these photos online, asking people to give him clues about what the devices were. A helpful person saw his inquiry and informed that he had in his possession the Guardian ST 820, a GPS type of device used for tracking. Afifi was alarmed when he found out that this type of device is exclusively created for and sold to the U.S. Army and law enforcement. Once the devices were no longer working, according to Afifi, the FBI retrieved them and told him that he did not need to worry or call a lawyer because he was "boring." This practice is a tool

WHO'S ON THE FBI'S MOST WANTED LIST? IT MIGHT BE YOU

You think you are surfing the Internet and browsing various topics in the privacy of your own home, but your digital feet are actually leaving big, bold tracks that can be recorded. In fact, if the FBI gets their way, your Internet service provider (ISP) will be required to track your browsing paths and store them for up to twenty-four months. This information would be available to local law enforcement as well as state and federal authorities if they serve your ISP with a search warrant or subpoena.

We were largely surprised by the lack of hue and outcry when the headline "Feds 'Pinged' Sprint GPS Data 8 Million Times over a Year" hit the news in December 2009. Perhaps people were too tired from dealing with bad economic news and planning for the holiday season's festivities to care.

But take a look behind that headline for a moment to see why you should be concerned. Sprint Nextel was asked by law enforcement through court orders and emergency orders to provide customer location data between the twelve months from September 2008 through October 2009, and they complied. In fact, Sprint Nextel provides law enforcement with its own self-service portal where they can set up automatic tracking for certain users based on their mobile phone numbers. That handy service on Sprint Nextel generated more than eight million transactions involving information sent to law enforcement. The fact is, if you are a Sprint Nextel customer, you may be under surveillance.

But Sprint is not the only carrier handling such requests—evidently all the wireless carriers receive about one hundred requests per week for customer-location data.

that the FBI and other law enforcement agencies use to protect our security, although since Afifi's discovery, the U.S. Supreme Court has ruled that this practice of tracking vehicles for long-term surveillance is not allowed without a warrant.

Is it only U.S. intelligence keeping tabs on their citizens? Not at all. In China, a woman posted a retweet on Twitter that landed her a year in a labor camp. The tweet was satire, but the Chinese government did not see the humor in retweeting a comment about smashing the Japan Pavilion that was erected as part of the expo 2010 Shanghai.[35]

The IRS Goes Online

Are you behind on your taxes? Are you fudging your income a little or a lot? If so, there's a new "friend" looking for you on Facebook and Myspace. You may recognize his initials: IRS.

The Internal Revenue Service (IRS) is combing through social media networks to catch disconnects between the income you might report on your tax returns and your lifestyle. The agency has a process that looks at items such as relocation information, professional profiles, and even postings that brag about expensive vacations.

Want an example? Here are just two out of many:

- The *Wall Street Journal* reported that the IRS Nebraska office found a DJ on Myspace bragging about spinning disks at a big party—income he never reported. Subsequently, they collected $2,000 from him.
- The IRS nabbed a Minnesota man for back taxes when he posted a comment on Myspace that said he was returning to his hometown to start a new job and then named his employer. The IRS swooped in and garnished his wages.

But as if government surveillance à la *1984* wasn't scary enough, there are other entities out there studying your online identity and actions.

Homeland Security

In 2010, ten alleged spies from Russia were deported. As their names, photos, and places of employment were announced on the nightly news, many people found that they were connected to the spies either directly or through a friend, or a friend of a friend, on LinkedIn, the professional networking website.

Why would spies join LinkedIn? The answer is simple: many of us take our trusting selves to the Internet. We link to people we know, they link to people they know, and then the circle widens. You begin by linking to people you met at a conference. You do not want to appear rude by ignoring their colleagues' requests to link to you. Then people join networking and information-sharing groups. Spies know about this behavior, and they listen, join, post, and assimilate information just like the rest of us. However, their purpose is to collect information for their sponsors, not to further their careers.

How could a spy lurking online affect you? You might be surprised. An alleged spy ring was busted in the summer of 2010.[36] One of the alleged spies for Russia, Anna Chapman, was fairly active on Facebook and LinkedIn.[37] While researching this book, we asked friends and associates to look at how close they came to being linked with her. One associate checked it out, skeptically, and then realized that several of his associates were connected to a person connected to Anna. Anna clearly knew who to connect to. One of her friends on Facebook was Steve Jurvetson. Steve is known for his venture capital work as a partner at Draper Fisher Jurvetson. When asked by the press about his connection to her, Steve Jurvetson replied in an e-mail, "I don't know her. So many randoms on Facebook." But think about it: In your career or personal life, what consequences could an online connection with a spy or a crook or a prostitute cause?

The Corporate World

Companies are watching employees for a variety of reasons. According to the 2011 Security Threat Report produced by Sophos, 57 percent of businesses are concerned that their staff shares too much information on the Internet. Employers are finding that it is effective to track employee surfing habits at work to protect their brand and security interests and make sure that nobody is disclosing sensitive company data.

The Society for Human Resource Management conducted a survey of companies and identified that roughly 75 percent have some form of Internet monitoring in place to track their employees. Employers track behaviors to make sure their employees are not breaking any laws by gambling or downloading pornography to the corporate network or individual computers. Companies may track your Internet usage, uploads/downloads of information or computer software, images that you look at online, and your incoming/outgoing e-mails and attachments.

Data Mining Your Life

Whether it's a creep or a company or a government looking at you online, every day they get a richer source of information about you as your online information piles up—and that record of your life is often permanent. Businesses are using the information you provide online in a variety of ways that might surprise you.

Social Exposure

Here's one example of a social website and how it treats your information. You might use Twitter to voice your displeasure with your company or even your boss. However, do you realize that your venting just became part of an online archive that is searchable by everyone? All public tweets from the time that Twitter began are archived at the Library of Congress. With the touch of a finger or a single mouse click, anybody can access a lifetime of posts, pictures, and secrets.

Exposure Both Unintentional and Intentional

Twitter isn't alone. Many of the services you use online, such as Amazon, eBay, Google, and Facebook, know a lot about you. Sometimes they share your details publicly for all to see, and sometimes they just sell that information to others.

Sometimes this exposure is intentional, as when Facebook shows high-level posts to people who search the site. Other times, these services share your information inadvertently, exposing it as the result of a massive technology glitch.

Here's an example of a site exposing you without intending to. Hackers targeted and gained access to *Gawker*, a New York–based blog focusing on celebrity gossip and other trendy topics. The hackers posted e-mail address information online, leaving many *Gawker* users exposed.

How Others Use Technology to Track You

Now that you have an idea about who's watching, we want to explore the technologies they're using to hunt down you and your information. From tracking software to geotags embedded in photos, these folks have plenty of tech help in keeping an eye on you.

Tracking: Harmful or Helpful?

When a website tracks your habits and your whereabouts via your phone or your computer, it can seem convenient. Technology can be used on a site to "remember" your favorite landing page. You can automatically log in to your favorite websites because your browser remembers your password for you. Sites can remember you and welcome you back, even making product recommendations based on your buying history.

Your computer may be able to track your online activities, but your cell phone may be able to track you as you move around the real world. If your phone was built for a cell provider in the United States after 2005, chances are you are carrying a snooping device in your pocket or on your belt. In that year the Federal Communications Commission required that cellular phone companies have 95 percent or more of the phones on their networks be traceable by satellite and other global positioning technologies (GPS). Your phone company, if your phone has a signal, can pinpoint your location to within roughly one hundred feet. Though this might sound creepy to you, this same technology helps to find lost or injured car drivers, helps parents locate their kids, and aids police in tracking kidnapping victims.

The current tracking technology is considered so effective that many shelters for battered women and children require that victims take their phone batteries out and store their phones, disassembled, so their abusers can't locate them using tracking technology.

When you can be tracked on your cell phone as a person of interest is still a little muddy. In the United States, law enforcement can track you on your phone. They can request assistance from phone companies (from January to June 2015, Verizon alone received more than 21,000 requests for cell phone logs) or use something called the Stingray.

In September 2015, the U.S. Department of Justice ruled that, except for urgent situations, federal agents must seek warrants before using a clone cell phone tower tool called a *Stingray*. The cloned cell phone tower mimics a legitimate cell phone tower and can record your phone's exact geographic location as well as numbers called while connected to the Stingray. As of the writing of this book, the Stingray itself cannot be used to eavesdrop on a conversation. Because it looks like a legitimate cell tower, you may never know you are connected to it.

"It's how we find killers. It's how we find kidnappers. It's how we find drug dealers. It's how we find missing children. It's how we find pedophiles," said FBI Director James Comey.[38]

TRACKING GOES TO WORK

Software companies are providing tools and analytics to help your employer track you when you are off the clock. Tools such as Social Sentry will tell your boss whenever you post something to Facebook, Twitter, and LinkedIn. The alert will happen even if you do not submit your posts at the office. The technology is advancing, and it has become easier than ever for your employers to keep tabs on you to protect their company's brand. Services such as InfoCheckUSA offer a service that provides social networking information on employees to employers.

At this time, if a state or local municipality in the United States wants to use a Stingray to collect data, the use of this technology does not require a warrant.

Law enforcement has other options at their disposal as well, but these often require a warrant. They can contact your cell carrier and ask for your GPS to be tracked. They can request your Internet protocol (IP) traffic from social media companies such as Instagram, Facebook, Twitter, or Google. Keep in mind that your IP address may reveal your physical location at the time you were connected.

Safe or Sinister: Checking into Trouble

There's a fun feature available these days from location check-in services that allow you to share your activities with your friends. You can use your smartphone, computer, or Web-enabled tablet to check in when you arrive at a coffee shop, business, shopping mall, or just about any place you can think of. This is a great way to announce to your friends where you are so they can find you. It also leaves you naked online, with your schedule, whereabouts, and habits available for all to see.

The check-in options are constantly multiplying, and in our first edition of this book, the popular services included Gowalla, Foursquare, and Facebook Places. The use of these services by businesses continues to morph. Today, you have many options to choose from including the three most popular: Foursquare, Meetup, and Yelp. By broadcasting where you are at all times, you also broadcast your habits and patterns to employers, friends, criminals, and stalkers. And remember, you are also broadcasting the locations where you are not, such as your place of employment, church, or expensive house.

But check-in can go wrong. A man was tweeting about a fabulous time he was having at a particular location. His friends played a practical joke on him and called the establishment and told them that his car had been stolen. The man ran out to the parking lot to find his car there, but the lesson stuck: When you tweet where you are, you are leaving yourself open to bad guys and friends with a bad sense of humor.

The authors were discussing this with a colleague and friend, a senior technology executive that asked to remain anonymous. This colleague was previously a human intelligence officer, and he specifically asks his colleagues never to post comments about how nice it was to see him until he has gotten home because he does not like to broadcast that his wife is home alone when he is away on a business trip. In a recent conversation with him, he advised, "Talk about where you have been, not where you are."[39]

Photo-Sharing Sites

Within six minutes of using a search engine and typing in somebody's name, we were able to access that person's profile and find her photo-sharing site.

Within those six minutes, based on how the woman had chronicled the lives of her children, we could tell what each child looked like, their names, family friends, recent vacations, and favorite activities. After we demonstrated this to a group of people, they were stunned at how easy it was to skim through the photo records of this family. There were enough details for them to guess ages, schools, and interests. In the wrong hands, the parents or kids could be targeted for ID theft or worse.

Photos with Geotags

Have you recently uploaded photos to the Internet? If you take a picture using your cell phone camera or a newer digital camera, it's likely that there are little codes inside the photo files that you cannot see but that tell a lot about the photo itself. Some of these codes help sharpen the image, but there are also geotags inside digital photos that include the exact coordinates where the photo was taken. If you upload photos the moment you take them, you are broadcasting your location.

A U.S. Army field artillery officer found out the hard way about geotagged photos. He was in Iraq and uploaded photos and videos to his favorite social networking and photo-sharing sites. He did not realize that by doing this, he was giving away the exact grid coordinates of the mission.

You may not be giving away military secrets, but if you take a picture of yourself or a loved one in front of your upscale house or shiny new car and upload it to the Internet, the information in the photos is broadcasting to everyone the latitude and longitude of your nice home and new car. Don't believe us? While conducting research for this book, we came across a website called pleaserobme .com dedicated to people who unwittingly invite criminals into their homes.

Cookies

Cookies are small files that computer programs put on your computer as you surf the Net. Sometimes cookies help with authentication, such as recognizing that you're using a personal rather than a public computer when you sign into online banking. Other cookies track items you place in an online shopping cart, even if you never complete the purchase. Many sites use cookies to customize your experience. A newer type of cookie, called a *persistent cookie*, stays on your computer even after you leave your Internet browsing session or turn the machine off. Most cookies are helpful and harmless, and you can delete them from your computer if they bother you, but in the wrong hands, they can be dangerous.

Ashkan Soltani, an online privacy consultant, conducted a review of fifty popular websites. These sites left a little present for each computer that searched the Net, including roughly sixty-four tracking technologies, some of which used cookies. This practice is what the techie world refers to as "cookie stuffing." Although cookie stuffing sounds like a fun way to bust your diet, this slang actually refers to the practice of stuffing cookies onto your computer without your full knowledge. Websites do this to peek at your online habits to target you in online marketing efforts. Consumers around the world, although resigned to this tracking, are still not thrilled with the idea that their personal online interactions can be monetized by companies for profit. In fact, in a recent survey a staggering 70 percent of Europeans have said that it is not fair that private sector companies profit from their online transactions and personal data.[40]

Most Internet browsers allow you to block or delete cookies and receive a notification before a cookie is installed on your computer. There are also several free browser plug-ins that you can use to add these features if your browser lacks them. One such service is called Ghostery, which works with several popular browsers, including Firefox, Chrome, and Internet Explorer.

Searching You Out

Just like the X-ray machine in the airport, search engines can quickly scan the contents of your life's baggage to reveal all kinds of facts in less than fifteen minutes.

In a test to show a family how easy it could be to mine their information, we searched for the mother's name, city, and church. We were led to a church bulletin, which told us about a charity project she was working on and gave her e-mail address. That made it easy to track down her Facebook account. The good news is her online persona was positive. Her digital billboard showed a health-conscious mom devoted to her family and church, a good neighbor, and a good friend. But no matter how good her online image, the search also revealed things about her identity that could be used to impersonate her.

Most Internet searches are purposeful and directed, using a specific name or topic, not random search terms. Posting thoughts, locations, and other tidbits about yourself online makes it easier for a stranger to guess your passwords or to target or impersonate you. Perhaps even more disconcerting is that most online snoops and scammers want to add to their existing knowledge of you over time, and they will patiently stalk and wait until the timing is right to take whatever kind of action will help them to take advantage of you.

Organizations, individuals, and criminals use Internet search engines to collect information and get clues about your interests and the interests of others you let surf the Net on your computer or smartphone. You might also find that the search engine providers themselves are using your searching information. On most search engines, every time you type a search term, you tell the search engine provider something about you. The search engine provider may know that you cannot sleep at night, you are interested in new diet fads, and you are considering getting a master's degree online.

In researching this book, we used a tool called Google Ads Preferences and were astonished by the stored knowledge Google had about us. Google knew that one of us was interested in national security, information security, and child safety. Google says that they are not tracking individuals, and they also offer a feature that allows you to opt out of their tracking.

Even though Google says it does not track individuals, they do track searches performed on your computer. They also store the searches for historical trending. So if you use someone's name plus other terms in a search, Google stores the combination of search terms. For example, if you searched "Jane Doe + Drunken Stupor," that search string remains in the Google archives.

Here's a tip: If the targeted tracking and historical hoarding by the search engines bothers you, check out the search engine DuckDuckGo, which promises not to track or gather information about you when you perform a search. Since the Edward Snowden revelations about digital data surveillance, DuckDuckGo has seen a 600 percent increase in searches using their engine.[41]

Public Records Exposed

You might be surprised by how much a stranger could learn about your family based on public records. If somebody is not sure if they're getting a good deal on a house, they can try the local public records office. The odds are their records are online. Your new neighbor can see what you paid for your house and get a look at your tax bill. If trolling through local public records sounds too cumbersome, aggregation sites such as Zillow.com do all the work for them.

If you have never "Google Mapped" your personal residence, take sixty seconds, go to www.google.com/maps, and type in your home address. Are you surprised at the stunning detail of your home? This service is not in real time (yet), but the quality, detail, and timeliness of the photos can be disconcerting. Google uses a mashup of satellite shots and maps and pictures they acquire by sending a Google Street View car or van down your actual street to snap portraits of your residence. This tool helps you get directions to someone's house complete with landmarks and photos to guide your way. It also gives crooks a good look at the best way to break into your house.

In the United States, these images are considered a matter of public record, and it is legal for Google to capture them. Where Google may have crossed the line was by picking up your wireless network in your house and recording it along with your home photo and geographic coordinates. Google has admitted to doing this and indicated that it was an accidental glitch in the technology. The company has since promised to take measures to stop collecting your wireless home network information.

Note that Google has provided a "report a problem" button on their website where you can ask for sensitive information regarding a property to be removed from the Google Maps and Street View. There have been heated debates in this case about the legal basis and protection of privacy for citizens in the United States, the United Kingdom, and other countries around the globe. Because of privacy concerns that could arise from this technology, 244,000 Germans said "no" to Google Street View maps by working through their court system.

ZoomInfo and Spokeo are sites that provide another example of how quickly someone can pull together information across the World Wide Web and see your life displayed on one screen in seconds. Spokeo calls itself the "white pages of people search" (we talk about the site in more detail in chapter 4). ZoomInfo bills itself as having the most comprehensive online profiles of companies and people. But, you say, nobody from Spokeo or ZoomInfo has ever asked you to fill out a survey. However, you actually do volunteer information every day through your online activities such as filling out surveys or setting up profiles on social networking and professional sites. Your offline activities, such as buying a house, buying a car, or paying taxes, also provide information that both services may be able to access.

Could Your Car Betray Your Identity?

There are all types of technologies in our cars these days. That technology helps us find our way, allows us to talk on the phone "hands-free," and controls the music that we listen to. Could technology in your car also be watching you and betraying your identity?

It is possible that the tech around you, much of which you do not see, could be keeping tabs on you. Given today's technology and the lack of laws regulating the use of the data collected about you, you should ask plenty of questions before you buy, lease, or rent a car.

The rental car company Hertz rolled out an on-board system that includes a fully featured GPS. Also included as a feature that is not currently turned on is one with full audio and video capabilities. In theory, this onboard system, if turned on, could spy on you by listening and watching you when you're in your car.

Here are some tips regarding your car and hi-tech exposure:

- *Protect your privacy.* You need to understand the authority of customer service agents. Do they have the authority to turn the audio and video on without an emergency call requesting the service?
- *Avoid hackers.* All technology is hackable and high-tech features in a car can create a potential entry point for the bad guys, especially if they want to target a high-profile person.

- *Ask questions.* When buying a new car or renting one, ask questions about the on-board system. Ask if you can disable or prevent snooping.
- *Understand legal protections.* In many countries, you have protections. For example, in the United States, the Federal Trade Commission has told businesses they cannot collect audio or video without warning customers.

CHAPTER 3

BEHAVIORAL TARGETING

[T]he fantastic advances in the field of electronic communication constitute a great danger to the privacy of the individual.

—Chief Justice Earl Warren in a 1963 Supreme Court opinion

W hen you use a free service, there is a quid pro quo. Many services can be offered to you for free because you give to them (or they take) information about you, a practice called *behavioral targeting*. They use that information to gain sponsors and advertising dollars or to improve their overall product to attract more users.

Some companies provide a free service but also offer a paid service with more options. This is sometimes called a *freemium*, in which customers love the free service and are willing to pay a little extra for the full-scale version or for in-service help. Think Candy Crush and the in-app purchases that help users move to a higher level. Both free services and freemium services probably involve behavioral tracking behind the scenes.

Have your kids or you used dictionary.com or Merriam-Webster.com to do research? Have you visited MSN.com or MSNBC.com to get the latest headlines? Those websites are free and helpful, but they expose you to high levels of behavioral tracking, according to a study.

In July 2010, the *Wall Street Journal* began a quest to understand behavioral targeting and how deeply it was ingrained into the Internet. They referred to it as the "business of spying on customers" and released an ongoing, informative series called "What They Know." You may have read parts or all of that series, or you may have read or seen other features about behavioral targeting in the news.[1]

Behavioral targeting has a positive side, but there is also a potential dark side that the *Wall Street Journal*, other media outlets, privacy advocates, and we are worried about—the loss of your privacy.

What's Going On?

There are many wonderful services on the Internet, many of them free, that you can access from your phone, tablet, gaming system, or computer. If you have an Internet connection and need to get the latest reviews for a new restaurant, find a doctor referral, or make a purchase, chances are you have used the Internet to do your research and help you make choices. What many of us do not realize is that added convenience comes at a price that goes beyond your Internet access charges. Behind those mouse clicks could be software that watches your activities and reports back to marketing firms.

Behavioral Advertising

This form of online surveillance is known as behavioral advertising or behavioral targeting. Behavioral targeting is often added to a company's existing marketing and customer service strategy to learn more about you so they can target ads to fit your preferences and sell things to you. They can watch your clicks, note how long you stay on a site, and how many times you have been on the site without making a purchase. Behavioral targeting also defines the tracking mechanism or technology that allows online advertisers, marketing firms, and other companies to sell to you more effectively.

Think of behavioral targeting as a secret agent who follows you around taking a small snapshot here and there of your Web browsing activities. The snapshot may include pages you visit, searches you make, and how long you stay on one site versus another. It may also "mashup" that information to create a best guess at your current location, the nearest WiFi hot spots, your gender and age, and even your unique device ID.

That information, fed to marketers, allows them to customize the ads displayed on websites that you visit.

Knowing that consumers are growing increasingly concerned about being tracked the Interactive Advertising Bureau announced a new initiative they call, "L.E.A.N.," short for Light, Encrypted, Ad Choice Supported, and Non-Invasive. The goal is to create an evolving and better set of guiding principles that consumers understand and agree to when creating ads that collect or track consumer behavior. Internet users want to opt out. It was recently reported that more than a quarter of U.S. Internet users use some type of ad blocker, such as AdBlock or AdBlock Plus. Even Apple has deployed ad blocking tools on their newest mobile devices.[2]

Data Tracking Trends

Behavioral targeting is a key component of the estimated $34.5 billion online advertising industry in the United States, according to a recent Juniper research report.[3]

Having companies track your activities is a mixed bag. Most people don't mind giving up some information about themselves for better service. Many of us let our grocery store track what we purchase each week by scanning store cards during checkout because it might translate into great coupons and savings at the checkout counter.

The trend for big Internet companies, such as Apple, Amazon, Facebook, or Google, is to move your preferences and data into computers under their control. This concept is sometimes known as the *cloud*. In the cloud your data might reside on both your computer and on a company's servers, or perhaps only on their servers. This can be a great way to offer you easy backups of your information, but it also takes away your personal control of how the data are stored and what they can be used for.

The collection of information on the Net is not as obvious, and you don't have a one-on-one relationship with online companies in which you provide your information knowing what you will get in return. So the first step in understanding your comfort level with behavioral targeting is understanding what is being tracked, how it's tracked, and by whom.

According to a *Wall Street Journal* study, in the United States the top fifty U.S. websites install an average of sixty-four pieces of tracking technology onto the computers of visitors. In most cases, the tracking happens without warning. Roughly two-thirds of those files were installed by 131 companies that make money off of tracking consumers.[4]

If you feel confident that you're fine because your browser blocks cookies, think again! The latest and greatest technology can track your Web browsing in real time. The tools can see when you access or browse a web page and then scoop up your device's location and can apply algorithms and other purchased data to infer about your income, health, and more. One example of these new tools is called *beacons*. Beacons track everything you do, down to where you move or hover your mouse on a web page. There are huge concerns about this pervasive tracking. For one, you as the user do not get an option to access the data and correct inaccuracies.

The element of big data analytics also plays a role. Many marketers claim to anonymize the tracking, but once you merge a few databases together the anonymized puzzle pieces form a full picture that can identify a person. In one *Wall Street Journal* study, a researcher indicated that computer programs can "de-anonymize" data if provided with thirty-three data elements.[5]

In 2014 the Federal Trade Commission (FTC) reported on nine data brokers that collect, compile, and sell vast amounts of personal information. One of the nine estimated that they have 3,000 data points on every U.S. consumer. They know all about you, from your shopping and personal habits to medical or health challenges and possibly even your political affiliations and beliefs. The

FTC indicated that most consumers have no way to check, correct, or confirm the information.[6]

A huge amount of public concern centers around government snooping and the bulk collection and storage of vast amounts of raw data in the name of national security. What you may not realize is that the greater threat to your privacy is often the collection of your data.

Danger: Data Brokers

Who is collecting your data? They're called *data brokers*. They can be research firms, Internet marketing companies, advertisers, and trade associations. The largest firms in this business are Acxiom, Epsilon, and Experian.

These firms can track, collect, collate, connect dots, and analyze your most confidential searches and your sensitive personal information. What do they do with the data? They package it up and sell it as a commodity. Oftentimes, they will anonymize your data but once your data are matched up with other

TIPS FOR MORE PRIVATE SURFING

As you surf online, the benefits of being tracked include being sent deals and bargains you might not find on your own. However, if you prefer to keep your searches more private, here are some tips that may help you. All of the tools listed here, unless otherwise noted, are free.

- Don't link across social media accounts.
- Treat your security questions and answers like passwords and don't share them.
- Use "Delete Me": https://www.abine.com/index.html.
- Use a tool called Privacy Badger that can block trackers: https://www.eff.org/privacybadger.
- Install ad and tracker blocker extensions in web browsers such as Adblock Plus or Ghostery.
- Take advantage of a virtual private network (VPN) such as Disconnect (disconnect.me, requires a small fee) to obscure your IP address location.
- There's a browser plugin called TrackMeNot that can turn your search engine histories into lists of unconnected terms.
- https://duckduckgo.com is a search engine that does not track you.
- Try "Google Incognito" to browse in private using Chrome.

databases for purchase, it can quickly lead back to your persona. This multibillion dollar industry operates behind the scenes of every website and mouse click or touchscreen swipe we make.

You may not mind filling out a marketing form to get coupons. There isn't anything nefarious about that, right? Over time, however, you are giving up private information online and you don't even know it. They can capture your likes and guess at your dislikes. They can see the connections to your closest friends.

What can these data brokers compile? Your religion, ethnicity, political affiliations, income, and family medical history, medications, alcoholism, depression, the bars you visit, and your purchases. As you click through the Web, third parties can grab your clicks and swipes. They can see what you read and what you do next. They may also grab the IP address and the device or computer ID which they can match up with other databases.

According to a *60 Minutes* interview, one company, Exact Data, offers to sell a database of names of people with a sexually transmitted disease or a database of people who have recently purchased sex toys.[7]

How Technology Comes In

In the physical world, many people are concerned about the level of surveillance they feel every time they go to a bank, shopping mall, or work. They see video cameras taping their every move.

You may not even mind being videotaped hundreds of times per day if it deters crime or helps law enforcement crack a case. But online you are not monitored by physical cameras; almost every website you visit on the Internet is tracking your online movements, not to prevent crime, but to monitor patterns and record your behavior to sell goods and services to you. When you are online you don't see cameras recording your moves because the technology is invisible.

Tracking happens on a variety of devices: when you're Web surfing on your computer, using your smartphone, gaming with an Internet gaming system, or swiping at the screen of a tablet such as the iPad. If you have a mobile device that talks to the Internet, just assume your activities are being tracked, even your local shopping mall wants to track you. During the 2011 holiday season, a mall in California and one in Virginia announced to shoppers that they would be tracking customer cell phones as the shoppers walked from store to store. They invited shoppers who wanted to avoid tracking to power off their phones.

How Are They Doing It?

Now you know what devices allow you to be tracked, but just how are they tracking? You are a key part of the problem or the solution based on what you do online each day. The first clue to how behavioral targeting works is to understand that you are broadcasting your thoughts, worries, purchases, and life through your online actions every day. Every click you make on the Internet might create a tracking opportunity based on what sites you visit, what those sites' information collection policies are, and what you do there.

When you visit one site, there may be ten or more companies looking over your shoulder and following you around. This is like walking into your favorite department store and, while you shop, having a sales clerk follow you around, noting your behavior. That sales clerk could also tap into information from the store's various departments—shoes, clothing, household goods, and so on—as you shop. "Female!" shouts the clothing department clerk. "Likes imported shoes," reports the shoe department. "Clearly in the upper-middle-class demographic from the handbag she carries, but also likes to rent romances on Netflix because she's got a movie sticking out of her purse," says the women's handbags associate.

If this kind of tracking and reporting happened in the physical world, you would either call the police to report this strange behavior or, at the very least, stop shopping at that store.

Behind the Scenes

Online, who is collecting these data is actually a complex question. It could be the federal government tracking you when you go to Whitehouse.gov or other federal government sites in an effort to give citizens better service. The tracker could be an ad agency trying to scrape up as much information as possible to add to their databases. Perhaps the company that owns the website is watching to help them understand their target markets. Even your own device could be "phoning home," just as E.T. did in the movies, to let the "mother ship" know where your phone and you are relative to nearby networks.

Behind the scenes we drop digital cookie crumbs. When you visit a website, unbeknownst to you, the website often asks for your IP address, which tells the tracking company your physical location while accessing that site. The website may note the language setting, the operating system of your device, time zone, and more. A great way to avoid this information being scooped up by advertisers is to clear your Web cache and consider trying an anonymous browser. For example, the Google Chrome browser offers an incognito mode and Firefox offers a private browsing option. You can also disable JavaScript, which prevents the collection of some of this information. Keep in mind that some sites will not work properly if you disable JavaScript.

Why You Should Care

Your response to the "what" and the "who" is tracking you at this point might be "who cares?" You might not care if a website follows you around to collect your viewing preferences to offer you better products or better search results. You have become accustomed to the Amazon.com model, which reminds you of what you searched for last time, recent purchases, and what purchases other people that are "like you" have made.

But before you decide to skip to the next chapter, you might want to read more about why this kind of tracking should give you pause.

Information Accumulates

The information tracked by online companies, both reputable and disreputable, aligned with open source information and public records, can make you naked online in a split second. We don't know what the future will bring, but the existence of this capability to track your behavior online is more than a little bit scary, and you should be on the alert for how it could have an impact on your online persona. In a *Wall Street Journal* article, a reporter talked about the work being conducted by a company called [x+1] Inc. (now known as Rocket Fuel).[8] You might not have heard about them, but chances are they know you so well you might just blush. In a test run by the *Wall Street Journal*, a volunteer submitted one click on a website. The company correctly identified the volunteer's gender, which was female. Of course, it's a 50–50 shot at getting that fact right, but the next set of data elements gave her pause: the software correctly identified a rough salary range, noted that she likes to shop at Walmart, and observed that she often rents children's videos.

Each person they tested had many similar details revealed in a spooky way. The technology continues to evolve. Tracking Internet visits down to a device ID is not the only consideration. Some behavioral tracking software can also record your keystrokes while you're surfing online and then send those clicks and screen flicks on your tablet or smartphone off to be consolidated and analyzed in minutes.

Follow the Money

One concern is that this snooping could be costing you dollars. For example, some travel sites modify search results based on device, purchase history, and search history, steering you to higher-priced search results or charging you more for room bookings. Orbitz uses your computer type to determine where to steer you. They previously steered customers who have a Mac to pricier rooms.[9]

Computer science researchers from Boston's Northeastern University proved that websites are tracking the online shopping habits of people and will charge individuals different prices, depending on what type of device they use to access a website. Their search results show that retailers such as Home Depot, Sears, JCPenney, Macy's, CheapTickets, Orbitz, Priceline, Expedia, and Travelocity charge higher prices to users on Apple or Android devices. Online retailers can and do adjust their prices based on your location, with higher-income areas receiving discounted prices.[10]

You Don't Have a Choice

Another reason you should be concerned about this pervasive tracking is that you, the user, don't get the option to access the data and correct inaccuracies. For example, Healthline is a company that helps marketing and advertising firms track people with health issues and then produce targeted ads for prescriptions and more. You have no way to correct or eliminate this information.

Understanding the Business Model

Companies using behavioral targeting have several objectives, but their main motive is to make money.

Your cell phone company tracks you because they need to know if you will retain their services or switch to another carrier. They also use tracking data to help build a better network of services. Social networking sites track you to offer demographic information to their business partners and sponsors so that they can sell to you and to be more competitive against other social networks. Many free sites track you to bolster their business case to their advertisers and sponsors or to sell your data to others.

What might be most surprising is the fact that a website owner may not even be aware of the tracking. After customers began asking about behavioral advertising and targeting, several companies decided they needed to double-check how their site handled customer visits. MSNBC.com and NBCUniversal began to monitor how much tracking was happening on their sites. The executives were astounded at the amount of tracking that happens based on the ads or sponsored companies they have included on their websites.

After doing an internal review, the *Huffington Post* also found similar tracking of behavior patterns. They were concerned enough to remove the ad firm Lotame Solutions, Inc., after they found that the firm was analyzing comments left on the *Huffington Post* website. Even Edmunds.com, the reputable source of anything automotive, found one of their ad providers snooping on their customers.

If this trend involved just a handful of companies, you might not be too concerned, but Krux Digital, Inc., looked at popular websites hosted in the United States and determined that almost one-third of tracking tools on those popular sites were installed by companies with access to that site and not the actual owner of the site. This might mean that your favorite sites have no idea of the extent of this tracking activity.[11]

When we used a tool called Lightbeam for research for this book, we found that in one minute and twenty-five seconds we had eighty-four third-party sites connected to us after visiting only six sites! We visited one shopping site, one social media site, and four news sites. At the end of three minutes of browsing the Web and doing normal tasks such as visiting Amazon and Facebook, we had 113 third-party sites tracking us and we had only visited nine sites.

How fast does it work? Try this quick test: turn off all your privacy tools and ad-blocking tools. Start at a favorite news site and notice the ads presented there. Next, go to your favorite search engine and search for boots. Go back to the news site and you are likely to see an ad for boots there.

Canvas Fingerprinting

Canvas fingerprinting allows a website to create a hidden line of text or a graphic that creates a digital token. It saves physical information about your computer, including your browser, screen resolution, operating system, and installed graphics hardware.

When the digital token is created, the way it gets rendered tells the website several things about you. Based on what they grab about you, your digital fingerprint becomes pretty unique. If they match the digital fingerprint with your search preferences, websites you visit, and more, they may be able to figure out who you are. Identifying you is a matter of math. The more data a company has about you, the easier it is to identify you.

How "unique" were we? In our case, when we looked up our canvas fingerprint on the site, https://www.browserleaks.com/canvas, the site said that

---

FEATURED TOOL: OPEN DATA PARTNERSHIP

Go to the Open Data Partnership at http://www.aboutads.info and use their "Opt Out from Behavioral Advertising" beta version (click *Consumer Choice Page* on the home page) to see the companies customizing ads for your computer. You can use tools here to opt out of receiving such ads or file a complaint. To use this site, you first need to enable cookies in your browser settings. After testing the beta site, we were presented with seventy-one companies that were participating in some form of tracking based on our device.

Seventeen of the companies were industriously customizing ads based on our past browsing behavior.

(continued)

The company names tracking us via the traffic on our laptop included the following:

AOL Advertising
Batanga Network
Casale Media
Clearspring Technologies, Inc.
Dedicated Media (DoubleClick)
eXelate
Google, Inc.
interclick, Inc.
Invite Media
Mediaplex
Microsoft Advertising
Quantcast
Specific Media LLC
Tumri, Inc.
Undertone
Vibrant Media, Inc.
Yahoo!

All of these companies were tracking us on a computer that is only used for work and has strong settings for reducing cookies and tracking. The work laptop used to test this tool is always set to "Only Allow Cookies for Sites I Regularly Visit."

The Open Data Partnership site's feature for opting out of tracking is easy to use. You can "Select all shown" to opt out of all the tracking and customized ads or you can pick and choose which to opt out of by the company name.

only 2 of 55,693 unique user agents, which you could consider as devices, had the same digital fingerprint as us.

How Your Behavior Can Be Used against You

Based on how you and your family use your devices, each click and site visit builds up a behavioral profile. Whether you like it or not, when your son visits

YouTube on your Android phone, your daughter visits shopping sites, and your uncle visits and wants to check lottery tickets on your home computer, they leave behind tracks that can be linked to a profile of you. Often the profile is tied to the device ID of the computer, tablet, or phone. Your Web activity patterns are then used to judge you.

Bad Customer

Those who run websites believe that, based on your behavior, they can estimate your future behavior and determine whether or not you would be a good customer. They may not offer you credit or free shipping because they don't think you will complete a sale or you could pose a potential credit risk.

Bad Behavior

If someone in your house likes to visit gambling or "adult" sites, this could create image issues for anyone who uses your computer to go online. Today's technology lets the behavioral tracking companies know that a specific device, such as your computer, tablet, or phone, has surfing preferences and habits. Although today's practices and technologies are not quite sophisticated enough to track your device behavior and fully integrate it into their decision-making process about you as a person or potential customer, we predict that in the near future, they will be able to. As the tools improve and directly tie browsing behavior to the device owner's name, you may find that you are denied a job or entrance to school because your child or brother used your computer for unsavory activities. We have not seen this happen yet but are observing a disturbing trend that could create scenarios where your device's browsing habits are considered a specific reflection of your Internet image.

CAN BEHAVIORAL TRACKING BE USED FOR GOOD?

Beside the ease of Internet use and convenience factor, tracking information can bring about good. For example, the technology behind behavioral tracking has been used to help in the aftermath of the tsunami and nuclear power plant disasters in Japan. You can look at the pictures listed by Dark Reading at https://www.theguardian.com/science/blog/2011/mar/24/fukushima -radiation-levels.

Price Discrimination

It is possible for a company, based on what it knows about your behavior, to target you and provide a custom offering based on the kind of person they think you are. This might not work in your favor if they offer you a deal that is more expensive than your neighbor's deal because they think you are an impulsive shopper or wealthy.

Social Engineering and Targeting

Behavioral tracking allows companies to collect data stores that may be combined with other data lists to build a profile of you. Tom Owad, a computer consultant, conducted an experiment that shows how these data stores could be misused. He did an initial review of the public wish lists on Amazon .com. These are handy lists where people can tag items they would like to receive as gifts. The lists include the name of the person and his or her city and state.

Tom downloaded more than 250,000 wish lists in a day. He spent time analyzing and consolidating the data he gathered from these lists and added them to addresses and phone numbers he found using Yahoo People Search. He was able to match much of these data with the Amazon list.[12]

How could such information be used? What types of books are on your wish list? Would you want to be located on a map by a group that opposes the point of view in the book that you are currently reading? Each piece of the puzzle, your likes and interests, could also be assembled to use in a social engineering attack on you.

Bad Citizen

Private companies love personal data, but according to the ACLU, so does the government, including the CIA and FBI. According to a news report from MyFoxDC.com, Chris Calabrese, counsel to the ACLU's Technology and Liberty Program, notes that "the government actually buys, subscribes to these databases and purchases access to these services."[13]

You may think privacy advocates are taking this issue too far, but we have heard more than one advocate express concern about what government and law enforcement can or might do with behavioral tracking data in the future. Would you get in trouble if you were on the Web researching gangs, listening to gang music, and then drove through a gang neighborhood roughly around the same time that a gang committed a crime while you had your location-tracking-enabled smartphone or tablet in your car? You left a lot of digital clues behind that could point to you, even though you did nothing wrong.

How Location and Behavior Tracking Work

When you land on a page that has behavioral tracking behind the scenes, the page uses software to scan information sent by your device. This information might include the device ID and/or the Internet address the computer uses. The software then takes that information and searches against other databases. One database might give them your computer's zip code. Another database may have guessed your gender and level of education. Another quick search of an information store may reveal a best guess for income and age based on the browsing behavior attached to that device ID.

TYPES OF COOKIES

There are many types of cookies, and new ones are being created every day, but here are a few you should be aware of:

- *Session Cookies:* This cookie is temporary. It follows you around one website. When you close your browser, this cookie goes away.
- *Persistent Cookies:* Also known as tracking cookies, these will track you all around the Internet and are resident even after you close your browser.
- *Secure Cookies:* If you visit a secure website whose URL starts off with "https," the cookie follows you around for the duration of a single Internet session, and it is encrypted.
- *Zombie* or *Respawning Cookies:* This sounds like a bad horror movie, and in practice, it can be. You delete these cookies, but after you do, they can re-create themselves on your computer. The most popular creators of zombie or respawning cookies are those cool media players on your computer that help you to play music or video, including Internet-based video. These are typically tied to your Flash Player and are sometimes known as *Flash cookies.*

In 2009, it was estimated that Flash software had been installed on roughly 98 percent of all personal computers. Flash is much more prevalent on non-Apple computers and mobile devices. Many media players use zombie cookies to make the player features operate, remembering your audio preferences, for example. However, these cookies also collect information about you.

Many sites use a small program called a *cookie*, which is downloaded to your device to identify you and track your actions whenever you visit the site.

In response to privacy concerns, Adobe created a feature called the "Global Privacy Settings Panel," where you can update your privacy settings. A good setting to look at is the "Always ask" before Flash Player allows a program to access your computer.

When you use the computer in your house, which has a unique device ID to connect to the Internet, cookies will keep a record of pages you visit, such as medical sites for a list of pediatric doctors, comparison shopping for diapers, searches for early learning educational entertainment videos, and family-oriented shopping sites. This behavior pattern might indicate that the primary user of the device is a parent.

It is difficult to pin all Internet traffic from a single device to a person with 100 percent certainty. However, this technology is evolving and is getting more accurate with each iteration.

The Facebook "Like" and Twitter "Tweet" Buttons

Most sites give you an opportunity to hit "Like" or "Tweet" when you land on their home page. If you are a big fan of the site, like their mission, or have had a great experience, chances are you clicked on one or both of those buttons. Little did you know that your action would allow somebody to track you.

It is estimated that the Like button for Facebook is on roughly 30 percent of the top one thousand most-popular Internet sites.[14] If you have logged onto Facebook or Twitter from a device and then started browsing around the

DOING THE MATH

Using a deanonymization algorithm, researchers can quickly and accurately identify you using pieces of the data puzzle to single you out. In one test reported by the *Wall Street Journal*, a man clicked on a website and the behavioral tracking software determined his location, education, rough income, and more. This provided enough information that a researcher stated it could be used to narrow down a search to sixty-four people in the world. The man's name was never provided, but having someone narrow results down to you and sixty-three other people on our entire planet must be an unnerving experience.

Internet, that site can collect your browsing information unless you log off. Sometimes the tracking is not being done by Facebook or Twitter but happens through shared apps such as Farmville that you may have added to your accounts. It is almost as if you invited someone to follow you around during your day to watch you, see what was on your mind, and follow your every move on the Internet.

Smartphones and Tablets

There are 5.3 billion mobile phone subscribers globally,[15] which is more than 75 percent of the world's population. The statistics do not really indicate if there are people that own more than one phone, but it is possible that some people have a smartphone provided to them by their employer and a personal phone. Depending upon the features of a device, a phone may be broadcasting its location and much more.

Not all phones are Web enabled, but consider that roughly 90 percent of all mobile subscriptions in the United Kingdom and Western Europe have Internet access. For a map view of how cell phone users in four cities around the globe use their phones, take a look at the http://www.manycities.org created as part of an MIT Lab Research project.

The most popular cell phone applications may startle you. They have changed quite a bit in recent years, with phone calls dropping to sixth place. Smartphones are now used first to send and receive texts, followed by reading e-mails, surfing the Web, and setting an alarm clock. They are also used to send e-mails, check the time, use calculator functions, and, coming in at tenth place was checking Facebook. While browsing the Web, smartphone or cell phone users enjoy reading the news, searching the Web, shopping online, and watching videos.[16]

In general, mobile device makers understand the need to act as better gate keepers and provide easier ways for users to opt in or opt out. Still there are dangers you should be aware of.

We Know Where You Are

Sometimes the behavior that's being tracked isn't online, it's where you go in the physical world. The Apple iPhone and Google Android phones regularly "phone home" their locations to company headquarters. Both companies are in a race against time to create the go-to source of information for pinpointing a person's location based on cell phone use and building an authoritative source of Internet WiFi hot spots.

This type of information can be used to create more accurate maps, provide helpful hints on how to get free Internet access, and even push notifications to locals about severe weather or other emergencies. But these conveniences come at a price: your privacy.

Google Android-based phones can beacon your data back to Google on the hour. Your device's unique identifier, location, and any nearby WiFi networks will be sent to Google. Apple also periodically collects data, including your iPad or iPhone's GPS location and any local WiFi networks. There is a small, unencrypted location file on these devices. The data are therefore vulnerable to hacking attempts. If you want to see what might be going on with your device, check out this tool at http://petewarden.github.com/iPhoneTracker to see your movements being mapped from your iPhone or iPad.

You Have a Spy in Your House

The research firm Gartner indicates that phone companies track you because they want to take advantage of the huge market for location-based services. Gartner estimates this market to be roughly $8.3 billion in 2014.

Phone companies may track you for a variety of reasons, including for the creation of WiFi hot spot databases. If you are a parent or employer, you may be happy to know that, if a phone is in your name, you can track your kids' or employees' movements based on their phone's location. (Check with your service provider for more details on how to do this.) The MobileMe service (www .mobileme.com) offers a find feature for locating iPhones (apparently when you have lost your device) that can be used to let you know where your kid's or employee's iPhone is at any time. This works because your phone often records every place you have visited. It records the address, the date, the time, and location. Now you know how they can find a lost or stolen phone.

Tracking capability isn't limited to smartphones. The MobileMe find feature also works on the iPad. An app called *MobiStealth* can be used to monitor users that have a Samsung Galaxy. Scientists have proven that once someone knows where you've been, they can likely predict where you will be going in the future.

The Data Exposed

The data that can be collected about you is extensive. Imagine that someone you don't know gets access to this data checklist:

- Device ID (which is unique and can be traced back to the contract holder, subscriber, or purchaser)
- Gender
- Income
- Age
- Marital status

- Number of people in the residence, including children and ages
- Zip code or other geographic indicators
- Personal interests, such as games you like to play or hobbies
- Past purchases
- Location history

Companies can use this data checklist to clearly target your needs and anticipate life events so they can offer you, for example, a car loan for that teenager who just turned sixteen. If the data checklist gets in the wrong hands, however, your identity can be exposed and you may become a target of a scam.

High-Risk Internet Activities

During his testimony before a congressional committee on June 18, 2009, Jeff Chester, Executive Director of the Center for Digital Democracy, warned that behavioral targeting (BT) is growing. A study by Datran Media, which surveyed more than three thousand industry executives from *Fortune* 1,000 brands and interactive agencies, found that "65% of marketers use or plan to use behavioral targeting. BT is expected to become widely used with online video, mobile phones, and online games and virtual worlds, further expanding its data collection and targeting role."[17]

Gaming

To see how far we have come since Chester's testimony, look no further than the companies behind the virtual world of gaming. One study showed that *Mass Effect 2* had 80 percent of their players using their sophisticated face customization program so they could create and change their avatars' appearances. *Mafia II* keeps track of how long your friend stares at *Playboy* centerfolds that are displayed midgame.[18] Did you know that this study of human behavior was happening? Maybe you or your loved ones would behave differently playing online games if you knew that anything you do during your sessions can be tracked, cataloged, and stored for future use.

Check-ins

Check-ins are a fun, somewhat cartoonish way to announce your presence on a social networking site. By checking in, you can access financial incentives or virtual rewards, such as coupons for free drinks or an award making you the designated mayor of the doughnut house.

There is a false sense of privacy among some avid check-in fans. When we asked several users of check-in about this, they typically responded that only their friends could see their whereabouts. However, one user named Jesper

Andersen found a privacy concern on Foursquare in 2010 when the check-in was accidentally broadcasting his whereabouts beyond his friends.

Social Networking

Social networking sites, such as Facebook, allow you to connect with old friends and keep up with current ones. They provide a great platform for friends and family members who are miles apart to feel as if they connected. However, these sites can put you at risk.

Every time you post something to a wall, click the "Like" button, play a game, or click on an ad, you are being followed. These services need that information to make their business model viable. That model might include profiling customers to design products that might do well in the marketplace, pushing ads at you to get you to buy something, or modifying their customer service message to appeal to the mainstream. In the case of Facebook, there are several documented cases in which user data, unknown to those users, have been cataloged and provided to third-party companies. One trick to watch out for is when new releases and features are announced that ask you to "opt in" to receive the full benefits of the new feature. Opting in can provide an open door for these services to violate your privacy.

If you are wondering how to reclaim your privacy, there is a neat tool at www.reclaimprivacy.org that you can use to scan your Facebook settings and get pointers to make your profile a little more secure.

But it's not just greedy capitalists who want to track your moves through behavioral marketing and targeting methods. There was a recent kidnapping case in which police believe one person's social networking activities made him a target. The kidnappers were actively using behavioral targeting on their victim's activities on social networks to learn about his daily routines and to plan their crime.

Retail Sites

Have you done any online shopping recently? If so, you probably weren't alone. Go to www.networkadvertising.org to see a list of who is looking over your shoulder. Entities that are directly related to retail sites you have visited watch what you do, even if you don't purchase anything.

CNN did a report on pricing customization and tested a retail shopping site.[19] They found that, based on the sites people had visited before shopping on the site, the person might find themselves presented with higher or lower pricing. One frequent shopper deleted his tags from his computer and noticed pricing actually went down on a DVD by roughly four dollars.

New retail models are emerging to respond to the privacy questions. Apple's mobile operating system iOS8, is said to contain a feature that sends random, phony MAC addresses to Wi-Fi networks to protect Apple users from being physically tracked through malls while they shop.[20]

Modifying Your Profile and Settings

Companies have started to create options for setting your profile in the "preferences" section of their sites. If you typically use Google or Yahoo!, you can take just a few minutes out of your day to check and change what information they have collected about your browsing habits. Other companies are also following the lead to allow you to see what they know about you. They typically offer you a profile, and you can review it to see what they think they know about you and edit it. This is a recent enhancement based on the general public expressing concerns over privacy. Many of these sites also give you the option of asking them to stop tracking you. Of course, they will probably still track you to a certain extent, but not at the same level of detail.

Change Your Ways on Google and Yahoo

We have highlighted the procedure for modifying your settings on two popular sites, Google and Yahoo!, so you can see how this works.

- *On Google:* Go to the Ads Preferences manager at https://www.google .com/settings/u/0/ads. You can add, remove, and edit your "interests" that Google has recorded. You can also ask the service to stop tracking you.
- *On Yahoo!:* When you open up your profile, it will tell you what Yahoo! thinks your computer's operating system is, which browser you use, and your interests. It will not allow you to add a new interest, but you can remove the ones they guessed at by going to https://help.yahoo.com/ kb/helpcentral.

Use the Tools

You can also leverage tools to help you manage how you are tracked and what information about you is gathered and passed back to marketers. Some tools to consider are Abine, Inc., at www.abine.com and TRUSTe at www .truste.com.

We do admit that fixing your behavioral profile on sites may feel like playing the whack-a-mole game. You fix one profile only to find that another company or service has created a profile about you that you need to edit, and sometimes that option is not even available to you. It may be worth your while, at least once a year, to review and track what is in your behavioral profile on the sites you use most often.

Here are some other tools and services to try:

- Browser plugin TrackMeNot can turn your search engine histories into lists of unconnected terms. You can also look at the options that NoScript (Firefox) and ScriptSafe (Chrome) provide.
- AdNauseum makes it hard for ads to track your identity.
- Google lets you clean up your profile and privacy settings all on one handy page: https://aboutme.google.com.
- You can ask for your personal profile as a marketing data summary from Epsilon for free. Go to http://www.epsilon.com/consumer -knowledge-center/marketing-data-summary-request.
- Try the Blur feature by Abine at https://www.abine.com/index.html.
- In the European Union, you have the "Right to be forgotten" (see chapter 8) where you can request information to be removed at: https:// support.google.com/legal/contact/lr_eudpa?product=websearch.
- The Ugly Email chrome browser extension can show you when your e-mails in Gmail are being tracked.
- Install ad and tracker blocker extensions in Web browsers such as Adblock Plus, the Privacy Badger from the Electronic Frontier Foundation, or Ghostery.
- Review how you "look" to others online at: https://aboutme.google.com.

Change Browser Settings

You don't need to be locked in to using Internet Explorer to browse the Web. There are several free Internet browser programs available, such as Firefox,

---

VENDOR FEATURE: GOOGLE

In response to the backlash by customers feeling like they were left exposed, Google has developed a central point to understand their privacy guidelines. They have also created a Google Family Safety Center to provide additional tips.

According to their website: "People have different privacy concerns. Our goal is to be clear about what information we collect, so that you can make meaningful choices about how it is used."

See http://www.google.com/intl/en/policies/privacy for the latest Google privacy policies.

Opera, Chrome, or Safari that provide similar or better features. The browser features change often, but most share certain functions you can use for privacy and security. Here are two important features to look for in your browser.

- Version: Keep the version of your browser up-to-date for the most advanced security and privacy setting options. You can check your version by clicking the Help menu and choosing About (Browser Name).
- Options or Preferences: Look for your browser's Options or Preferences feature to change its security and privacy settings. Most browsers will allow you to decide when and if cookies are installed on your computer and allow you to delete some or all cookies currently on your system. You may also be able to opt out of ads or pop-ups.

You can also go to the site networkadvertising.org and use their tools to opt out of targeted ads. Keep in mind that sites remember to avoid sending you targeted ads by installing a cookie on your computer to keep track. So you now have a tracking cookie to help fulfill your request not to be tracked by cookies.

Four Tips for Protecting Privacy

Here are four important areas you should address to provide the basic tools for protecting your online privacy:

1. *Apps:* Only download apps that you need to have and make sure the source of the mobile smartphone application (called *apps* by the industry) is reputable. Only turn on location tracking services when absolutely necessary.
2. *Data Usage:* Monitor data usage to see if the mobile device sends out data even while you're sleeping; this could be a sign that an app is broadcasting your location and other information.
3. *Settings:* Check your privacy and location settings to make sure they are set at the level with which you are comfortable.
4. *Operating System:* Stay up-to-date on the latest operating system updates so you have the latest features for privacy.

CHAPTER 4
SELF-EXAMINATION

O ver time, each of us develops an Internet persona. Sometimes the digital image of your life is accurate. Sometimes the image highlights one aspect of our lives above others, offering a true, yet incomplete picture. Occasionally, an image is a complete fabrication that can have harmful consequences.

How can you tell what you are exposing online and what online persona you've created? You must take the time to find out what is being written about you and what images of you are flashing across the world. Have you been stripped bare by your own online revelations or overexposed by someone you thought was a friend? You can't understand or modify the world's perception of you until you've done some online sleuthing. Scouring the Web for information about yourself is the first step to understanding and managing your online image.

What Can Go Wrong: A Case Study

Former *NBC Nightly News* weekend anchor John Seigenthaler read a disturbing Internet report about his father that was totally false. His father, also named John Seigenthaler, was a prominent journalist and founder of a successful public relations firm who once served as Robert Kennedy's administrative assistant. Seigenthaler Sr. had even been a pallbearer at Robert Kennedy's funeral.[1]

The Seigenthalers were therefore shocked to read an entry for Seigenthaler Sr. in the online encyclopedia site Wikipedia that stated that Seigenthaler Sr. "was thought to have been directly involved in the Kennedy assassinations of both John, and his brother, Bobby." The article went on to incorrectly report that "John Seigenthaler moved to the Soviet Union in 1971, and returned to the United States in 1984."[2] The same false reports were repeated on the websites Answers.com and Reference.com.

It is impossible to know how many people read these reports and developed a false opinion of the Seigenthaler family as a result. According to an article he wrote in *USA Today* in 2005, the senior Mr. Seigenthaler referred to these false reports as "an Internet character assassination."[3]

These false statements were available on certain Wikipedia sites for 132 days until the Seigenthalers discovered them and requested that they be removed. Seigenthaler Sr. did not create or post these lies online. He did nothing to earn or publicize the false aspects of his online persona, and he was able to correct the misstatements relatively quickly on discovery. However, any person researching him or his family on Wikipedia during those 132 days would have received completely false and scurrilous information and would have taken away an entirely incorrect impression of the man.

Wikipedia is a site that many people have used for basic research, treating the information there as undisputed fact. For the Seigenthaler family, sophisticated professional managers of media and public image, Wikipedia became an unchecked medium for propagating harmful lies and damaging one man's reputation.

John Seigenthaler Sr. has written extensively about the false postings on Wikipedia and about his attempts to find the people responsible for defaming him. His interviews on the topic for CNN and National Public Radio demonstrated the dark side of how the Internet can be misused as a research tool, and it highlighted how quickly libelous statements can streak across the Internet.

The Seigenthalers corrected the online lies, but here's the point of this story: the family would not have been able to address this problem if they didn't know it existed. They were fortunate that a friend notified them about the false Wikipedia entry. But it's not wise to wait for your friends to tell you about online damage to your reputation. Everybody should conduct a regular inventory of his or her online persona.

Where Are You Now?

In today's social media saturated world, many people's Internet persona is a product of their own posting, liking, hash tagging, and uploading. If you have a favorite social media site, then you have probably posted significant information about yourself, and anyone with access to the site would also be able to see your friends and family, your relationship status, and many aspects of your personality. There are likely to be pictures and video as well, either posted by you, or tagging you.

So take a close look at the entire history of your own postings. First confirm that your site is managed on a friends-only setting, so that you can limit and control the number and type of people who can see what you post and whose comments can appear on your page. If your social media site is open to the public, think about what would happen if important people in your life saw the content of your postings. Would their review of your publications affect your life in a negative way? If so, then flip the switch and make your site private, or clean it up so the worst postings—bad temper, admissions of drug use or alcohol abuse, sexual statements, off-color or hateful jokes—are no longer online and attributable to you. We all slip sometimes and post things that we would regret later or that didn't come out the way we intended. The current rules of social media sites always allow you to remove your own posts, so that you will not be stuck with your mistake forever. So take advantage of the rules and scrub down your social media presence.

Then do it again for any other social media companies whose services you use. Also remember to close out old accounts if you are not using them anymore. And keep in mind that even if you remove a picture from your social media feed, the same picture may have been copied by another person, or it may remain available to the Internet for a period of time in the memories of search engines like Google or Bing.

Surprising Sources

You may think that the only information about you online comes from postings on your own Facebook page, but you might be surprised what information you find when you check yourself out online. You might just find more than you bargained for.

Snapshots from Friends

Although Internet references about you may not be in your complete control, you can still influence those who post them. For example, if your friends or family include unflattering or embarrassing pictures of you on their social media sites, blogs, or personal Web pages, you can ask them to remove the pictures.

Most friends will understand if you want to remove that picture of you at a bar, hoisting an appletini high in the air, if the friends know that you are about to apply for a job and the company human resources director is likely to be looking for information about you on the Internet.

Friends, family, coworkers, and neighbors may also be posting comments about you online. Some of these comments may be hosted on your own social media page, but many will be viewable on others' pages or the pages of your mutual friends. Many people are willing to help you clean up your online persona when you express concern about the comments. Keep your friends close.

Big Brother

Most, if not all, of the contacts you have made with the government are recorded somewhere, and many of those records are available online. Some

WIKI WHAT?

So what is Wikipedia, and why can it be used to spread false information? Wikipedia was started in 2001 as a free, collaborative online encyclopedia. The site's primary rule is to strive for a neutral point of view in its treatment of all subject matter. The nonprofit Wikimedia Foundation operates the site, and it does not accept commercial advertising. Wikipedia claims to have an accuracy rate similar to the *Encyclopedia Britannica*, and there is no question that the online format allows for a timely, updated description of a topic that matches a fast-moving, information-based society.

In 2011, Google published a list of the most-visited websites in the world, excluding adult sites and Google's own. The fifth most-visited site on the list was Wikipedia. With more than 18 million articles at this writing (3.4 million articles in English), Wikipedia is both revered and reviled as an online resource. Both the strength and the weakness of Wikipedia arise from its unique structure as a collaborative resource. The term *wiki* describes a website built on special software allowing contributors to add to the website's content using an Internet browser.

The site can provide immediate and often comprehensive information about subjects as diverse as professional cyclists or obscure mountain peaks, modern politicians or ancient cities, movie quotes or named hurricanes. However, although some Wikipedia articles are seriously researched and written by professional historians, geologists, or other trustworthy academic sources, the vast store of information is written by non-experts or people interested in pushing a particular agenda. In addition, although Wikipedia has a staff of editors and it responds to complaints of inaccuracy, Wikipedia articles are not always closely reviewed and edited. Therefore, while millions of entries are entirely accurate and complete, Wikipedia is typically not treated as a trusted resource for serious academic study.

of these government databases are password protected or they otherwise limit access, so a search for your name would not list the information they contain about you. However, people who know what they are looking for can unearth a vast trove of personal data from government websites.

If you ever started a company, served as a director of a nonprofit organization, or reserved a business name in your state, then information about you is probably accessible at the website of the Secretary of State. The same site would have records about you if you ran for office, lobbied state legislators, or became a notary public. If you have registered a car, secured your commercial driver's license, or ordered personalized license plates, then your information is likely to be searchable at your state's department of motor vehicles. If you run a restaurant, your sanitation scores may be posted on the website for the local county inspector's office.

Keep in mind that, no matter what you do, some government information is completely beyond your control to affect or change in any way. For example, if you are a homeowner, your county may choose to list all home purchase prices and tax valuations online. You are unlikely to convince the county to stop displaying the price of your house unless you can demonstrate that there's a serious error in the data.

It's a Crime

If you've been convicted of a crime, you may be covered by different rules. Hamilton County, Ohio, contains the city of Cincinnati and publishes a list of deadbeat parents who have been held in contempt of court or have been convicted of nonsupport for failing to meet their child support obligations. The Hamilton County sheriff's office allows full Internet searches or browsing through a list of names and addresses and includes the type of official ruling made against each person. At the time of this writing, the list has more than 2,500 members. According to the sheriff's website, all people on the list are classified as "wanted" and could be arrested on sight by law enforcement. Not all counties make it this easy to find the names and addresses of child support truants, but the federal child support enforcement laws make lists of habitual scofflaws available to law enforcement, who, in turn, may publicize the lists.

Of course, official criminal records are also moving online. Registered and convicted sex offenders may desperately desire to separate records of their convictions from Internet searches for their names, but most states will continue to publish these data, no matter how damaging it may be to the former convict's current reputation. It is likely that you can quickly find a map of all of the convicted sex offenders in your neighborhood and review the details of their convictions.

State prisons and jails in larger counties usually post their "inmate lookup" tool online so you can find the status of your loved one quickly and easily. California offers sex offender searches and information in many languages. A public nudity conviction would clearly land you a place on your county's website (and probably on the local newspaper and television station websites as well). The Florida attorney general lists the ten most-wanted criminals on its website, along with press releases discussing successful prosecutions. You can find public records online for people who have been convicted of crimes, those who have been arrested in sting operations, and individuals who are currently being pursued by police but may not yet have been arrested or convicted.

Although the government posts official records, private groups and individuals aggregate the public data into further lists. For example, an organization called the Violence Policy Center keeps lists of people with licenses to carry concealed handguns who were convicted of killing with those same guns. The site includes the names of killers and victims, the locations and descriptions of the crimes, and a listing of mass shootings carried out by concealed gun permit holders. It is likely that we will see more of this type of website as the Internet matures and organizations use the vast store of public data to make political or social points or to argue for changes in the law. Once your name is listed on a public record, there may be no limit to the number of ways that information is displayed online.

Victims of Crime

Victims of crime are often shielded from having their names displayed, but certain categories of victims, such as the missing and the murdered, are more likely to appear online. Various states post missing-persons directories that often show missing people's names and the circumstances of each person's disappearance. We all hope that information about us or our families does not appear in these online databases, but if it does, it is available for anyone to see.

Your Day in Court

Any attempt to explore your Internet persona must include a search for all court decisions to which you were a party, and even those in which you may have appeared as a witness. Interacting with these public records may show you a new side of yourself.

Of all three branches of government in the United States (executive, legislative, and judicial), the judicial branch is the one most likely to host your information or list your name. You could appear in the public record for testifying at a legislative hearing or petitioning your state's public utilities commission, but your name is more likely to be made public if you are involved with the courts in some way. This is true even assuming that you've never had contact with the criminal courts, have never been arrested, or have never even been pulled over for a traffic violation.

Millions of marriages in the United States end in divorce, and the records of divorce and family law proceedings are often posted online. If a family law matter is settled or remains decided at the lower court without appeal, then the records are likely to be public but only available for personal viewing at the county clerk of court's office. However, if one side appeals the decision, then family law court of appeals decisions are much more likely to appear online, although often not in easily searchable formats.

State Supreme Court decisions are even more likely to appear online. If you have been involved with a divorce matter that was appealed, you should visit the website for the relevant county court system and search for your records so that you can know what others could find out about you. As you might imagine, divorce cases can be packed with emotional outbursts and the financial details and demands of both husband and wife. All these details can be found in many published court decisions.

Civil cases arising from automobile accidents, deals gone bad, business battles, or product liability can also produce case decisions that appear online and describe the situations and even the personalities of the parties involved in each case.

For example, in the 1996 English libel case of *Berkoff v. Burchill*, the defendant, a film critic, had repeatedly called the plaintiff film director ugly and was sued for saying so. The court allowed the suit to proceed to conclusion, finding that calling a person "hideous-looking" could hurt his self-esteem and his earning potential. The entire case is reported in many places online and is searchable, which is likely to cause more problems for the reputation of the plaintiff than the original remarks. Court cases show us at our most vulnerable—vain, greedy, angry, and frustrated—and online court decisions may be the worst possible additions to our online images, but it's important that you know that they are public records.

Databases and Organizational Records

Consider that some sites on the Internet exist to collect information on people to publish for the public good, or to sell. These sites have not asked your permission to use your name and the information about you.

To find this information, you should think about what actions you have taken that are likely to be collected in a public database or held for consideration by a wide group of people. Some database companies learn about you and your family for their profit. They collect databases that you cannot find by running an Internet search on one of the major search engines but that contain deep and personal information about you and your family.

A perfect example of this type of data-heavy website is Ancestry.com. Ancestry.com calls itself the world's largest history resource and is one of dozens of genealogy sites that profess to help people find their roots by learning as much as possible about their parents, grandparents, and generations before. As you might expect, Ancestry.com provides databases of marriage, birth, and death records from multiple countries to allow you to trace your family line back through time. It also includes census records and voter lists going back into the 1790s, immigration records, passport lists and ship passenger lists from before the U.S. Civil War, and family trees for reference. All of these tools can take you back in time and help you to find your family, but they also include a database of school yearbooks, where you can find pictures of yourself, your siblings, or even your mother in second grade or graduating high school.

These database sites featuring genealogy, military history, regional information, or any collection of useful information can be used to find out more about you and your family. These sites are growing as they collect more information from books and records, building them into databases and making them searchable online. Some of these sites are slickly professional like Ancestry.com, and others are collected by amateur historians.

In fact, every group that you ever joined, from the Campfire Girls to the YWCA, to the Daughters of the American Revolution, may have a record of your participation, and more organizations' records are moving online every year. Your workplace may have a biographical page describing your talents to the world. Your church, temple, mosque, or ashram may list you in the membership or leadership rolls and show pictures of you and your world-famous potato salad at the last potluck dinner event. Every group membership that a researcher finds online tells more about you.

Some group memberships reveal more than others. Some affiliations speak to your level of wealth or social class, whereas others tell people about your ethnic background. Others will provide information about your children, spouse, or pets. You could be a member of a singles club, a belly-dancing class, or a trivia team, all with an online presence or pictures from their latest event. If you are active in your labor union, Internet viewers know about your work life. Your high school or college could be congratulating you for your promotion at work in their online alumni newsletter or thanking you for leading the twentieth

OFFLINE DANGERS

You may be happy that people can easily learn about your club memberships or religious affiliations. But even if you don't mind people knowing about your memberships, you still may not want a casual researcher to know where to find you next Wednesday night because they see you are a regular player at the senior center's bingo night. Expressing your affiliations online can lead people to find you in real life.

anniversary fund-raising campaign. The more you do, the more of your life is probably reflected in the website databases of your favorite organizations.

Exposure by Shopping

Think about what you bought this week. You may have bought diapers and baby food. Maybe you bought oil filters and fan belts to work on your car or purchased dress patterns or a book about dealing with depression.

Anyone who is watching our purchasing habits can learn a great deal about our lives. It is unlikely that you would regularly buy diapers and baby food if you did not have a baby at home. An observer can infer what model car you drive by the auto parts you buy. And it would not take the deductive powers of Sherlock Holmes to conclude that a person who bought a book on depression is coping with this condition in himself or herself or somebody near to him or her.

When we discuss revelation of your secrets through shopping, we are not describing information that is available to the general public. While a nosey neighbor and a prospective employer may be able to read your public government data, your social media, and run a Google Images search to see what pictures have been tagged with your name, they will not have access to your shopping information. So your purchases may not provide clues to your personality to just anyone.

However, all types of retailers and other marketing companies, including search firms like Google, grab and keep this information, and they use it to supplement a profile of you that was created long ago. There are also companies called *data aggregators* whose entire business is comprised of collecting information about people and selling that information. For data aggregators, sometimes your data are sold directly connected to your name or your home, and sometimes they are bundled with the data of many people who have similar traits to yours. For example, a restaurant many want a list of people who have Googled Mexican restaurants in the past six months in your town so it could send them a coupon. Or a political party may want a list of men who own pick-up trucks so it can send them a solicitation for donations.

Finally, a crook who has installed snooping software on your computer can determine your shopping habits and use those data to his advantage. Special spyware can follow your movements online and send it to whoever installed the software onto your system. If you buy an expensive item and have it shipped to your house, the crook could be waiting for it.

How Your Shopping Habits Reveal You

You are revealing a great deal about your activities by committing your hard-earned resources to a product or service and exposing information about your lifestyle through what you buy. And most people have the items they buy online sent to their homes, handing their addresses to companies and possibly thieves.

Why would anyone want to learn about you from your shopping habits? In fact, such knowledge is the goal of many organizations. When you purchase a book or music from Amazon.com, the company is building a profile of you based on your buying habits, and this profile allows them to target you with advertising that they believe is more likely to spur you to buy additional items. As stores and advertisers grow more sophisticated, they merge the information they know about your online buying behavior with your life and purchases in the physical world. If they can predict your behavior based on past purchases, then maybe they can influence future purchases.

Who's Looking?

Consider the trail of sales information you are leaving on the Internet. When you buy shoes online, it may not only be the shoe store that notes the transaction but also the bank that holds your credit card and the managers of other sites you may have visited that are tracking your movements online. Any of these companies may be passing on the information (a) that you are willing to purchase items online, (b) how much you are willing to spend, (c) your method of payment, (d) the exact nature and category of the item that you bought (including shoe size), (e) what type of online store you are visiting (electronic boutique or Amazon-like superstore), and (f) what sites you visited before and after your purchase. Although some of the same information may be collected and shared from a regular shoe purchase at a store in the mall, it is not as easy to process and not available to as many different parties.

Given this knowledge, if there are products and services that you want to acquire but that you do not want associated with your name, then you are better off not purchasing them online. Although online shopping may be more anonymous in certain ways—you do not have to physically expose yourself by walking into an embarrassing store or parking your car in the lot—your Internet movements and purchases are more easily monitored and recorded without your knowledge or consent. Internet purchases are tied to your online access account, your e-mail account, the special number that identifies the computer device you are using, and ultimately can be tied to your name. You should keep this accountability in mind when deciding how and where to spend money.

In addition, you should think about the items and services you purchased online. The last trip to Florida you planned at Priceline.com and that kitty condo delivered from the Internet PetSmart may be adding to your online persona. Take account of who might have learned of your purchases online and what they might say about you. You can influence how much data are available about you online by choosing how and when to spend your money.

Of course, as marketers and retailers learn more about how to collect your information online, they develop an insatiable appetite for information collected from every corner of your shopping life. Always remember that when you receive a discount, your shopping information may be paying for that discount. Why do you think that your local grocery store has a VIP program where they scan your card for a discount on your purchases? They are capturing and using that information, and they may even be selling it to third-party data aggregators. Who would buy diapers unless they had a baby at home? People with babies are susceptible for buying other baby-related (or maybe sleep-related) items.

So aside from these information gathering devices that you use yourself—like the grocery VIP card—stores are finding new ways to capture data about you out here in the nondigital world by using digital technologies. For example, digitization allows a store to install cameras that not only capture your image, but also use facial recognition technology to follow you around the store to see where you spend the most time, and to tie your face to your name when you use your credit card at the checkout counter. Then these data are combined with data the store holds from your online shopping habits, providing a better picture of your purchasing habits.

Some stores take this even further by pinging your smartphone when you are in the store, and not only combining these data with the facial recognition data, but following your smartphone to the next store you visit, and recognizing you the minute you return to the original store. Digital shopper tracking has started on the Internet, but it has become so valuable that retailers are finding ways to mimic the online shopping data capture at their brick-and-mortar stores.

Data Mining for Fun and Profit

It's not just retailers who are collecting information about your online habits. If you think about it, so much of what you use online is free—search engines, mortgage calculators, news services, and more—that you have to wonder how these sites make their money. Once they have gathered a herd, how do they milk it? The answer for many of these sites is data mining. Also known by industry terms *big data* and *data analytics*, data mining involves gathering information and selling it to others, and it's big business, with Google as perhaps its largest online practitioner.

In a 2010 *Time* magazine story, writer Joel Stein called a number of data mining companies that were stealthily collecting information about him "taken from the websites I look at, the stuff I buy, my Facebook photos, my warranty cards, my customer-reward cards, the songs I listen to online, surveys I was guilted into filling out and magazines I subscribe to."[4] Stein found that many companies were collecting this information and making assumptions about him, some accurate and others inaccurate. One company pegged Stein and his wife as liking gardening, fashion, home decorating, and exercise, whereas others noted that he rents sports cars and buys intimate apparel.

Stein notes that there is now a "multi-billion-dollar industry based on the collection and sale of this personal and behavioral data" and that each of these pieces of information is sold for about two-fifths of a cent to online advertisers. The information is linked to his browser while he is visiting sites online, and that information is used to display advertisements that are most likely to entice him to click through to a purchasing page and buy a targeted product or service. He notes how creepy this online knowledge stalking can be, writing that "right after I e-mailed a friend in Texas that I might be coming to town, a suggestion for a restaurant in Houston popped up as a one-line all-text ad above my Gmail inbox."[5]

Data analytics has only grown in importance for businesses since Stein wrote that article in 2010. Two years later, the *Harvard Business Review* was calling the data scientist "the sexiest job of the twenty-first century."[6] And by 2015, Google's senior vice president of global marketing, Lorraine Twohill, is quoted by the consulting firm McKinsey and Company saying, "The tools available to marketers today are extraordinary. They know far more about their consumers than ever before. They are able to have a much more meaningful, two-way conversation. It's definitely the golden age for marketing in many ways."[7]

As observed throughout this book, capture of your information and finding meaningful ways to use it, combine it with other data, and learn more about you is becoming more deeply sophisticated each year, as retailers and marketers learn from the way they used your data in the past. You can safely assume that all the data tracking described by Stein in 2010 is happening now in a way that is even more useful to the marketers and probably more intrusive to your privacy. A "golden age" for marketing means an unprecedented amount of information on consumers is being collected.

Every piece of information about you has value to somebody and adds to the online image of who you are and what you do.

Searching for Yourself

When asked where to look on the Internet if you want to reveal secrets about someone, Miami resident Vincent Volpi of the Private Investigation Company of America (PICA), an agency with twenty-four fully staffed, international regional headquarters and more than three hundred correspondents in cities worldwide, stated:

> I'd look for media articles in all languages and court cases in all venues where the person has lived or done business. There are no better sources than spurned

lovers, wives, business partners or competitors. Everything has to be taken and presented in context and verified, but these are where the secrets lie and will be talked about. The media stories would be somewhat self-validating. Court testimony as well.

Volpi said that investigators build on what they find in media and court records by talking to the people involved. Civil, domestic, and criminal courts can provide detailed information about our worst moments and failings, and court records of all sorts are becoming easier to search and find online.

Where to Start

When asked where his investigators would begin exploring online to find information about a person of interest, Volpi said, "We would run the routine searches that anyone sophisticated with the Internet would run and we might utilize some of the commercially available sites depending upon where those inquiries led us. There are many services, such as Intelius, that harvest quite a bit of information from generic sources and can lead you to places where you can use human intelligence gathering to get more valid facts. Also, there are many proprietary consolidators of information like LexisNexis that, when creatively used, can be a big help. So can PACeR. So can various services that are offered through the Net where someone actually goes out and hand-searches court records." For deep searches, according to Volpi, PICA investigators can start with these cost-effective Web tools, then "resort more to Humint [human intelligent gathering] and contacts and old-fashioned 'gumshoe' work."

You can identify who you are online by using tools that professional investigators rely on every day and tools that you use as you browse the Web.

Your next step is to use those tools to begin to build a profile of your online persona.

Following the Trail You Left

When exploring your online persona, visit any sites that you have published online. Ask yourself whether you:

- have your own website, your own Twitter profile, or your own professional biography page through work;
- are linked into others on LinkedIn;
- have posted your face on Facebook;
- have an online journal or a place where you publish your deepest, most emotional thoughts;
- are contributing recipes to Cooks.com;
- have left a profile on eHarmony or another online dating site;

- sell handmade purses or landscape paintings on eBay or run a commercial site where people can buy your goods;
- are leaving comments on sports pages, political discussions, or local government websites; each of these online activities leaves a trail, so you should make a list of any and all such sites.

When you have found and listed all the websites where you have posted information about yourself, analyze the depth of your participation in each site. How much information about you is included in each site? Examine whether you are participating actively in the site, making groups of friends and entering into public chats, or if you only listed a small "billboard" about yourself and took no further action.

Make a special note of those Internet places where you are the most revealing about yourself. You could be giving away information about yourself every day on a site through your friendships, your comments, your pictures, or your open dialogues. If you are blogging or producing an online journal, then you are likely to be giving people access to the most important facts of your life and your daily thoughts, frustrations, and desires. Think about the picture that would be painted of you if someone connected these sites together, learning about your emotional life and frustrations from your online journal, your workplace priorities from LinkedIn, and the identity of your friends and family through Facebook.

Going Visual

Uploading multimedia can be especially revealing. Pictures and videos show the people you are with, the parties you attend, the clothes you wear, and the places you hang out. Family reunion shots expose all of your relatives, the warmth (or coolness) of the interactions between you, and often the details of a relative's house and its physical location. Automobiles included in photographs can say much about your personality—whether you are a Corvette person, an electric car person, or a minivan person—how much you spend on a car and in what condition you maintain it. Photos at work can show your ID badge with company identification numbers, and shots of your house can reveal your street address. Pictures of you at the beach generally show more skin than you might be comfortable exposing to coworkers.

Exposing Likes and Dislikes

Sometimes it's *not* what you say or what you show that uncovers you, but what you enjoy. Think about what products, people, publications, videos, organizations, or jokes you have claimed to "like" online.

BODY LANGUAGE

A recording of you singing a karaoke song can reveal many things about you: it's one thing to read a printed profile of a person and an entirely different thing to see that person, to hear her, and to watch her laugh and move. Social scientists have repeatedly shown how much information, both intellectual and emotional, humans glean from seeing the expression of another human.

Dr. Paul Ekman of the University of California at San Francisco started the Diogenes Project to document how much information we can learn from reading faces. He has tracked the forty-three facial expressions that humans can make and what each expression says about what we are thinking.* Looking into a face we see information about health and happiness, about taste and fashion, about expression and emotion. With a picture, we have an opportunity to read information that the subject did not even know she was revealing. The Web is a visual and auditory medium, allowing us to make those judgments about people we see online. How much of you is showing?

---

* Malcolm Gladwell, "The Naked Face: Can You Read People's Thoughts Just by Looking at Them?," *Annals of Psychology*, August 5, 2002, viewed on Gladwell.com.

Many social networking sites provide countless opportunities for you to pass jokes or pictures on to your friends and to register your approval of the latest trend or dance or YouTube video. Content aggregation sites that provide Web-surfing accounts such as Tumblr or StumbleUpon can reveal much about your personality by showing a stream of the websites and pictures that you find appealing online, and under their default privacy settings, nearly anyone can see your preferences.

Certain music sites operate the same way. You can discover much about a person's personality by knowing whether they prefer mainstream country music to 1940s swing or obscure modal religious chanting. You can learn even more if you know what photographs or videos appeal to that person. Pay attention to the messages you are sending to online observers based on the preferences you divulge.

It's in the Mail

Correspondence is another medium for leaving an impression online. E-mail, private messaging on social media sites, text messaging, chat, and other online correspondence provide clues about you. However, given the nature of this type of messaging—one-to-one and often fleeting—it is not usually worth scouring the Internet to find and retrieve these electronic breadcrumbs. Much of this information, like text messages or many online chats, is not saved. Nearly all your correspondence is intended for a single person, and that person will either delete it or save it, but probably never share it, so it is highly unlikely to sneak into your public Internet persona.

If your friends post your e-mails or text messages online or if your messages are mostly posted in public on others' walls in social media spaces, then you should take these into account when you review what you have posted online.

Keep in mind that the person who receives your correspondence may not be the only reader or the only one who decides what is saved and what is sent into oblivion. E-mails sent to a business address may be held on a company's server for years, and today nearly every company has reserved for itself the legal right to review their employees' e-mail. If a business is sued, it may be required to save all e-mail entering or leaving its server and turn the e-mail over to lawyers for review. Some families use computer control software that allows one computer in the home network to see all correspondence that passes through the network, even e-mail or text that seems to have been deleted.

Gmail from Google uses computer algorithms to analyze the text of messages sent from or into its service. The company likely attaches the information about you to their sophisticated marketing profile of you, and more immediately, provides advertisements on your e-mail page that are influenced by what Google thinks you are talking about in your messages. So even if your electronic message is written to one person and that person deletes it from his or her inbox immediately upon reading it, someone else may know what was in the message, and your message may become part of what people know about you through the Internet.

Delving into Databases

A good exploration of your online persona should include digging deeper into the database sites that relate to your life. These include your school data, your city's history, religious information, and data about your early work life. Skipping across the top of the World Wide Web can take you only so far. People researching your life will dig deeper. Many public databases and informational websites are used as a starting point by professional investigators trying to learn about a target.

Hiding Behind a Torn Screen: Flaws in Internet Anonymity

People sometimes use pseudonyms or "handles" to identify themselves in comments at interactive sites, whereas others sign up for dating sites or social sites such as Myspace or Xanga using false names or nicknames like "lonelygirl16," "evilclown," or "dogsrool." When you follow the trails that you have left online, assume that, whatever name you've used, a researcher can trace each comment, photo, or item of personal information back to you.

Using a false name online can shield you from being exposed as the person who always makes nasty comments about cat lovers or the woman looking for love in Phoenix. However, you can't be certain that a false name will protect your identity. There are many ways that your true identity could be discovered.

Tech Tools to Help Find You

There are Internet tools that others can use to break through your barrier of false names and find the many facets of your Internet presence.

Social media companies may not know your real name, but before you can comment or add information to the running discussions online, you must set up an account using your e-mail address. The research site called Spokeo allows reverse lookup of your e-mail address. Using reverse lookup, anyone can identify your "handle" on social media sites. Spokeo scours the Web for social sites connected to your e-mail address and displays comments, pictures, videos, and preferences in user accounts connected to your e-mail address, plus the "handle" that you used to enter this information.

Other tools that you'll find at sites such as emailFinder.com or Lookup emailaddresses.com perform similar functions. New tools are added to the Internet all the time, and people can use them to discover your online pseudonyms and tie your real name to the comments you have been leaving online. These are also great tools for you to use to build a profile of your online presence.

But Spokeo is not the only tool to unmask Internet users who thought that they were anonymous. *Forbes* magazine describes an entire class of online marketing companies that charge retailers to track down the identities of anonymous website visitors. Companies such as Relead, Visual Visitor, and Loopfuse charge for the service of generating more targeted leads by better identifying the people who visit a website.[8] And remember that Google and other search firms are keeping vast amounts of information about you and may be able to provide that data to retailers on request. Marketers are tracking your movements online, and if some of those movements involve a registered site used under an assumed name, then your assumed name just becomes one more point of data in your retail file.

Even the FBI is finding new and creative ways to find people who believe they are anonymous in their use of the Internet. One of the methods of carefully protecting against exposure of your identity online is to use it through specialized free software and an open network like the one called *Tor*, that protects against all kinds of traffic analysis software. *Wired* magazine reported that the U.S. FBI has adopted a well-known hacker's tool called *Metasploit* and its "Decloaking Engine" to "stage its first known effort to successfully identify a multitude of suspects hiding behind the Tor anonymity network."[9] Metasploit is described by *Wired* as "the most important tool in the hacking world: An open-source Swiss Army knife of hacks that puts the latest exploits in the hands of anyone who's interested."[10] As we might expect, law enforcement is using available tools to track down suspects, but these tools are available to others too.

And always remember that lawyers in the midst of litigation have important tools at their disposal. (See the It's Legal box that follows.) A subpoena served to your Internet service provider (ISP) can yield important information about who you are and where online you have been. Most ISPs will not give up information about you without some type of legal process document, like a subpoena or a court order. Learning your IP address, the numbered address of any website you are controlling, can lead to a lookup of your name and contact information on the "Whois" tool available to anyone. So there are a number of tools for someone who really wants to hunt you down and break through even the most carefully planned online anonymity.

Learning about Yourself Using Search Engines

The World Wide Web contains billions of pages of information that are publicly available to anyone with an Internet connection and a browser. Millions of pages are added every day, which could make it challenging to find any needle in this enormous haystack.

To keep all these data discoverable, as information flows onto the public Web, much of it is captured and cataloged by search engines. All day and night, search companies send out "spider" programs, crawling and cataloging the Web. The spiders find new sites and note the changes to existing sites. With this huge database of Web information, search companies then apply software that finds links and connections between and among various points on the Net, determining popularity and interest levels, eventually ranking Web locations on various different criteria. When a query is typed, the Internet search company uses complicated calculations to guess what each searcher wishes to find on the Internet from the search terms entered and then to propose likely targets of each search. That makes search engines handy tools you can use to your advantage in discovering your online persona.

Looking for You

Internet searches are not limited to finding e-commerce or informational sites. You can search for yourself to see how many sites contain a mention of your name and whether pictures or videos display your unique charms on the Web.

IT'S LEGAL

When anonymous speech crosses a legal line into defamation or unfair business practices, the legal system of the United States provides a way to uncover the identity of the speaker. The victim of the unlawful speech can petition a court to subpoena records that display the identity of the speaker. How does this work? Every person enters the Internet from some device. Your home desktop computer, your laptop from work, or your mobile smartphone all rely on an Internet provider to access the Web. Each Internet provider has a specific numerical identifier that pinpoints Internet traffic originating from their service. It's likely that your provider has also assigned your computer or smartphone a number so that all of your Web surfing, comments, uploads, and downloads can be traced directly to your device. Although there is software that can mask this process, it's not always perfect, and the majority of Internet users don't use technology tools to stay anonymous. Most of us rely on using a different name for various accounts, never realizing that such behavior will not affect how comments can be traced through Internet protocols.

In normal circumstances, companies, individuals, or governments will not have access to the information that traces your comments back to your Internet account. But someone who convinces a court to issue a subpoena or a valid court order is likely to gain access to the data, assuming that the relevant websites and Internet providers have kept the information on file. The anonymity of standard Internet users is no match for the legal system. This is an important way that your identity may be compromised online.

Searching for your name on Google, Yahoo!, or Bing is known as an "ego search" because many people do it just to stroke their own ego by finding out who is talking about them and measuring their popularity or success by the size of their Internet footprints. This seems to be the modern-day equivalent of looking to see if you are mentioned in the local newspaper's society page.

But searching for your name on the Internet is also an important step in discovering and managing your online persona. You need to see the depth and character of information about you and what other people can find out about you online to know if the information is accurate and to repair any serious image problems.

Keep in mind these tips for constructive searches:

- Use quotation marks around your name. If you search your name without enclosing it in quotation marks, the search engine may provide early results that not only identify you, but it will also uncover sites that use either your first or last name, finding other people, places, or things that share your first name or last name. The quotation marks instruct the search engine that you are only interested in matching all the words listed in the order in which you typed them.

- Seek out all major variations of your name. If you are a married woman, Web information about you may be listed under your married name and under your maiden name, so check both. If your name is spelled differently in English or is often written with the appropriate accent marks, then search for both versions. Remember that millions of websites are written in English, but millions are likely to exist in your home language as well.

- Check the formal version of your name and any nicknames that people might use to identify you. The background check for "Fast Eddie Labeque" is likely to unearth a different type of Web reference than the search for "Edward Labeque," "J. Edward Labeque," or "Jackson Edward Labeque." Try all variations for the broadest possible results.

- Include all titles or honorary terms that might apply to your name, such as "Dr. Edward Labeque" or "Edward Labeque Jr." The purpose of this exercise is to discover as many references to yourself as possible on the Internet so that you can see an accurate picture of your online persona.

- Search with important words from your life in addition to your name. The search for "Tracey Smith" is likely to produce scores of pages on various people named Tracey Smith. Sorting among the mass of irrelevant Web pages is time-consuming and frustrating. Break through the clutter by including the city you live in, your workplace, or spouse's name in your search. These modifiers are more likely to pull information about you to the forefront. In other words, once you have found everything possible for "Edward Labeque," search the Web by attaching "Galveston" or "teacher" to the name, or search "Edward and Catherine Labeque." Frequently, this will pull targeted sites up toward the top of the search pages and make them easier to find.

- Include words that highlight the reason you might have been featured on a Web page. Look for articles that address your sporting victories by adding the words *tennis*, *rugby*, or *skiing* to your name as appropriate. If you sell real estate, try searching with the word *realtor* next to your name.

- Use several different search engines, and use their various tools for a broader range of results. When using Google, first try the traditional Web search to find any sites that mention your name. Then proceed to the Google Images search, a Google video search, and YouTube videos search to see what media content is posted about you. If you are an academic and want to find your work, Google has a special search engine precisely for this purpose. Use it to search both your name and your publication topics. If you are a journalist seeking your work, also try the name of the publication you work for, and search in Google News.

- Keep in mind that images captured by search engines are not always pictures of you but may be pictures of people who are related to you personally or professionally. Images may include books you have written or products that you have endorsed. For example, Google Research captures articles that may have been written by you or about you. We are always surprised at how often our names appear in Google Blogs, capturing forgotten interviews and articles, despite the fact that neither of us has a blog of our own.

- Review as many search engine results pages as possible. The item you want may be on page 16 or page 2 of the results, so you may have to scroll through lots of pages. And you never know when an apparently obscure picture noted on the tenth page of a Google search will gain popularity and move up to the first page. Sometimes this happens because the reference itself receives links or hits from other sites. Often results move in rankings because the site containing your picture or the content around your picture moves in the rankings for no reason related to you. Either way, it is best to run a broad and deep search to learn as much as possible about your online persona.

Search engines are not infallible, and they cannot read your mind. Even the most sophisticated search tools only recognize the precise phrases you type in the search box, often influenced by the types of searches you have run previously on the same computer. So experiment with terms and learn about how to broaden or narrow your searches. Start broad and narrow as necessary to weed out unwanted data. Starting with a broader search allows you to see if any information exists on your topic, rather than to try to capture everything online about your topic in a series of narrower searches. Reducing scope as you proceed can facilitate your inquiry into a variety of related topics: "Edward Labeque dentist," "Edward Labeque scrabble," "Edward Labeque charity," and "Edward

Labeque grand theft auto" are all searches that will bring out varying aspects of Dr. Labeque's life.

Search engines are only as useful as the information they return, which is driven by the search algorithm the company uses and the number and types of sites searched. All search engines have significant limitations. The Google engine prioritizes information based on the concept that the more sites link to this information, the more important it must be. This premise has led to many fruitful searches, but it may not find all significant references to you online.

Google's site notes that you are looking for answers, not trillions of Web pages, so the Google algorithms are designed to provide the answers you are most likely looking for. Google states that "Today Google's algorithms rely on more than 200 unique signals or 'clues' that make it possible to guess what you might really be looking for. These signals include things like the terms on websites, the freshness of content, your region and PageRank."[11] For this reason, Google will guess that you are seeking the "China Dynasty" restaurant in the town you are living in, rather than a restaurant in a different town. And the evolving Google presentation screen will provide you with the phone number, menu, Web site, operating hours, and location of that local restaurant because studies show that these are the facts you are most likely to be seeking by initiating your search.

Google searches information listed on public sites posted on the World Wide Web. Much information about individuals is listed in databases that may be accessed through the Internet but will not turn up in a standard Google search. Using other search engines or search tools like Google Research or the

FAME AND OBSCURITY

The challenge of searching for yourself can be more difficult if you share a name with a celebrity. A *Time* article about Internet name searches noted that if your name is Brian Jones and you're not the former Rolling Stones guitarist, then you don't exist on the Internet. Of course, the problem has nothing to do with your existence and everything to do with the extensive publicity given to celebrities online. Fortunately, major search engines provide a way for you to exclude certain terms from your searches. You're much more likely to find the Brian Jones you seek if you type the following phrase into the search bar: "Brian Jones"-"rolling stones"-guitar

Yahoo! Directory may help but may still leave online information about you undiscovered. Search engines are helpful and will find information that you otherwise would never have known about, but they are not a panacea for finding all references to you. This is why we shouldn't stop with searching for people using major online tools. You are likely to find information in many more places.

Analyzing Your Profile

As you search for yourself online, highlight or save links to Web sites that contain information about you (a tool such as Microsoft OneNote can be a great way to catalog these links). You can also copy the URLs (Web addresses) to a Word document or write URLs down on a sheet of paper.

Once you have pulled together all the information about yourself that you can find online, it's time to examine the results. Taken as a whole, your online persona is likely to show a distorted image of who you are, leaving out important aspects of your life. You should think about whether this view of you is telling the world too much or too little. Should it be pared back so there is less data online about you, or should it be supplemented so that people will have a more accurate picture of you?

Start with an overview of your entire collection of data. If someone was able to find all of the information you just uncovered, what would he know about you? What impression would he have of your personality, your life, and your

FEATURED TOOL: SPOKEO

There is a set of data aggregators that collects information both online and in the physical world to sell to investigators, employers, curious mothers-in-law, and anyone else who may want to learn about you.

One site that offers a uniquely Internet-centric set of information is Spokeo. Started by Stanford students as a way to aggregate their friends' social network postings, Spokeo evolved into something much more insidious. Spokeo searches dozens of picture, video, music, and social sites and finds all information tied to a single e-mail address. This means that anyone who has received an e-mail from you, or who knows your e-mail address, can find many of the accounts that you opened using the e-mail address. Even if you think your Twitter feed, your Instagram account, your dating profile, your

(continued)

political blog, or your Shutterfly picture albums are anonymous, a Spokeo user who has your e-mail address will be able to easily tell that you are the name behind the account.

Spokeo calls this type of search a "reverse e-mail lookup." When you enter a person's e-mail address in the tool, Spokeo provides you with a summary that may include the name and probably the address of the person attached to that e-mail address. If you are a paid member of Spokeo, you could see the various online profiles of that person; the person's participation in social networking, dating, music, video, and online shopping sites; all information that Spokeo can find on the person's family; data on the person's house, including photographs and neighborhood evaluations; wealth and income data; lifestyle and personal interests; and authentication that reveals whether the e-mail is actively in use. Looking up a person by name or phone number can also provide political affiliation, race, highest level of education, religion, and reading material.

Spokeo uses a search tool that works differently from those offered by other search engines and data aggregation sites. The Spokeo website states:

> Spokeo can help you learn about people who have shaped our world by providing one of the largest libraries of online famous people profiles. Browse photos and videos and timelines, including the public whereabouts of your favorite stars. Research a new neighborhood before you move through Spokeo's reverse address search or lookup any people who will be in close contact with you and your family.

Spokeo claims to harvest information from phone directories, government databases, social networks, mailing lists, and business information sites. Although all of this information may be publicly available, it's much easier to pull together in one place using a tool such as Spokeo.

Joining a data aggregation site like Spokeo will shortcut your search for online information about yourself. These sites pull together full profiles based on deep digging for public data and save you clicking through dozens of sites to find your online profile. Many of these sites show financial and property information. Spokeo, in fact, searches and displays a total of nine categories of social networking sites.

habits? Would he or she know where to find you on a given day and what to say to you so that you would like and trust him or her? Think about the types of people who might be looking at your online persona. Would a nosey neighbor find out things that you wouldn't want him or her to know? Would a prospective employer or serious boyfriend or girlfriend find information that paints you in a bad light or that would hurt your future prospects? Would a potential robber or identity thief find enough data to pretend he or she was you at the bank or to convince you of a scam?

Consider conducting an exercise in persona analysis. After you've reviewed all the information you collected about your online persona, construct a full picture of the character described there. Based only on the online data, who is this person, and what are his or her priorities in life? Do you like him or her? Could you track him or her down in real life using only the information you collected online? Fill a page or two with descriptions and analysis of the person you have just researched. Include a paragraph of missing items—important current or historical personal facts that you could not find online. Using this detailed analysis, you're ready to decide which items should remain online, which should be removed, and what should be added to present the identity you want to own.

Revealing Yourself Online: A Checklist

Building a full profile of your online image helps you control what everyone else sees when they search for you. The following checklist should help you complete this task.

- Check the trail you have blazed on the Internet: Social media profiles, online journals, dating sites, photo collections, and video sites
- Location-based sites (like Foursquare)
- Music sites
- Do these sites allow you to limit access (friends only)?
- Have you signed up using a fake name?
- Have your friends and family posted information about you?
- What presence have you left for others to see? Shopping sites, merchant sales (do you sell products on eBay or other sites?), preferences (publicly displayed likes, dislikes, and comments), and correspondence exposed to everyone?
- Is there information revealed by search engines: Google, Bing, Yahoo!, and Ask.com. How many pages of results did you find for each search? Have you used specialized searches such as for images or video? Have you run a deeper search on sites such as Spokeo?
- What information is available on third-party databases, in newspapers, and other media?
- Have you checked government agencies such as courts for records?
- What other sites could hold and display information about you, for example history or genealogy sites?

While it may be frustrating or depressing to find embarrassing information about yourself online, you are better off knowing what your online image is. As you will see in the next chapter, there are several strategies available to you for removing or obscuring content from sites on the Internet. Knowing the extent of the damage is the crucial first step to understanding how to fix it.

CHAPTER 5

TIME TO GET DRESSED

Now that you have researched and analyzed your online persona, it's time to repair any damage to your reputation on the Internet and create a new, better online image. Part of this exercise is reactive—cleaning up the problems that you discovered and limiting access to those places that should be kept private. Part of the exercise is proactive—building and creating a presence on sites that people will use to judge your character.

This chapter describes the tools you can use and actions you can take to manage your Internet persona. We examine broad strategies for adjusting your image, including the care and cleaning of your online profiles and sites. We tell you how to remove old or embarrassing content from the sites of friends and family. We also explain what's involved in getting information taken down from business and government sites. In addition, you get an overview of various legal remedies that may help cleanse your Internet persona and of the limitations of these remedies.

Choosing Change

Have you ever driven down the highway and thought about what each vehicle says about its driver? Whether you drive a flashy sports car or a pickup truck you choose to project a certain image, just as you do when making choices about what to post online. It's time to go car shopping, and choose the exact online image you want to display.

Before we dive into remedies, we want to encourage you to give some thought to the new online you. This time around, you should gain control of what information goes out there, putting yourself forward in a positive way and incorporating new information that shows you in the best possible light.

You must first decide what you want your image to be. If your life is changing drastically—for example, if you are graduating from school and entering the workforce—it is likely that your more youthful online persona along with childish resentments and pranks should change to reflect what you want others to see. If you move to a new city for a fresh start, why should you drag your old habits and friends along with you? If you are about to be married, maybe it's time to remove all those dating site accounts. Changes such as moving to a different town, expanding your family, taking a new job, all shift your life in ways that should be reflected in your online persona.

One of the authors of this book attended a wedding reception where the siblings of the bride and groom read out loud to the guests the online dating advertisements that attracted the happy couple to each other in the first place. This "reading of the eHarmony profiles" ceremony made clear the changing personas of the bride and groom from "single and looking" to "committed and happy." Many characteristics of your old life can remain online long after you have outgrown your connection with them. Clearing out these obsolete signposts is an important step in establishing that you are committed and happy with the changes in your life.

Human growth entails recognizing each new age and building on the past, and the Internet can help us grow. Even if you want to advertise your current life and preferences to everyone, you will probably want to update your advertisements on a regular basis. No one wants to be judged on old data, and no one wants to be thought of as the person she used to be.

Setting Goals

If you've realized that, for whatever reason, it's time to make a change, it's a good idea to first set some goals. First, consider the size of your presence online: your Internet footprint. Knowing whether you want a broader, deeper Internet persona or a more cultivated and curated, pared-back version will help you to set priorities and make decisions as you build your new Internet identity. So many choices now exist to help you project your online image that your first selection between blossoming and pruning will provide an initial direction.

Perhaps your goal is to minimize your online footprint. You may want to strip back how much people can see about you online, closing down accounts and making others available to "friends only." Are you embarrassed by the pictures and comments on your Facebook or Instagram page? Then limit the number of people who can see them, and keep them hidden from employers and future dates. Have you moved past an old relationship, an old career, or an old haircut? Then pull those aspects of your former life down from the Web entirely. Clipping your Internet presence doesn't have to lead to a tiny Web presence. It can be the start of a process to build your Internet persona, this time from the solid base of carefully selected content. With this approach you can keep a low profile on the Internet, or you can use your new smaller presence as a foundation for building your new Internet persona.

Alternatively, you may want to keep your Internet presence as it is with few changes but manage it more effectively. Maybe you like the information you find when you Google yourself, and it serves as a terrific starting point for a more self-directed footprint. Or maybe you find your Internet presence to be perfect right now. If you like what you see, your job is to simply keep your image current and clip the weeds back every now and then.

You might decide to maximize your online footprint, diving into social sites such as Facebook, Buzznet, LinkedIn, and Twitter to help you network for that next job or relationship. You might flood the Web with your opinions, ideas, and activities, creating an expanding and complicated online persona that replaces your older image with new energy. You might dive deeply into every website you can find that supports or comments on your favorite topic—Eighteenth Century British poetry, consumer privacy, Warner Brothers cartoons, stock prices of the Fortune 500—and demonstrate your interest and expertise within a community narrow in scope, but broad in geography.

There are as many Web persona strategies as there are human personalities and experience. Whatever strategy you choose, at least you will become aware of what people can find out about you online and you will have taken steps to manage your image so that it reflects your current priorities.

It's a Lifestyle

Your Internet persona is a living, growing thing that can sprout new roots and branches before your eyes, even if you do nothing. The search you perform in six months will probably turn up different items than the one you perform right now, and the growth in your image captured by the Internet may not be the growth you want people to see. So occasional and relentless online hygiene is an important aspect to protecting your reputation.

Rethinking and amending your online image is an exercise that you may want to consider repeating regularly. Like spring cleaning or replanting a garden, you can set aside time once a year to analyze the ways that your image has changed on the Internet, pruning the unsightly growth and adding some fresh, new data to reflect any changes in your life.

Remember that technology, social trends, and the information others post about you will constantly change and you'll need to monitor how this change affects you. Think about the fact that a social media site like Facebook didn't even exist until 2003, and didn't gain widespread popularity until much later. Entire worlds have grown online in that time, and it is likely that new worlds will emerge and expand over the next several years. The Internet continues to grow as millions of people add more information on current sites and more types of new sites come along. Some of that information will describe you or highlight aspects of your behavior or your personality.

STRATEGIC THINKING FROM A PROFESSIONAL

For an expert perspective on cleansing your online persona, we talked to professional reputation consultant Henry Fawell, president of Campfire Communications, a strategic communications firm in Baltimore, Maryland. Henry advised,

Your online reputation matters. It's not static. It's a handshake. It allows customers, journalists, employers and investors to size you up, look you in the digital eye, and make judgments about your character without ever meeting you. The question is not whether those judgments will be made; the question is whether you will actively seek to influence them. We need to treat our online reputations like our property. When our lawn gets overgrown, we mow it. When the pigeons do injustice to our car, we wash it. When our suit gets wrinkled, we dry clean it. Why would we treat our reputations any differently?

Your online persona is here to stay, and the care and maintenance of this image should be as automatic and routine as updating your résumé when you go on a job interview or checking yourself in the mirror before you head out on a date.

Keeping Your Private Life Private

Before the Internet our lives were much more private. We had to make an effort to expose our lives to public view and scrutiny. If you wanted a broad group of people to know about your exploits, you would have to have found a reporter so you might be seen in the newspaper, heard on the radio, or both seen and heard on television.

Now, in the age of social media and behavioral information collecting, our lives are public, and we have to make an effort to keep them private. Pictures of you may be published by many people and seen by hundreds, thousands, or if you go viral, millions.

In this age some people accept having their entire lives viewed by others. But many of us believe that every last thought, opinion, picture, and connection should not be available for viewing online. Many believe that only they and those

EVEN SPIES HAVE PERSONAS

Some people are actually put at physical risk by public exposure. British government official Sir John Sawers was selected in 2010 to be the leader of the British Secret Intelligence Service, known as MI6. Before he could take that office, the press noticed that his wife, Lady Shelley Sawers, maintained a Facebook page full of pictures and intimate details about her family. As stated by one press site, "This means that information such as the names, photos, and whereabouts of the couple's children, the apartment the couple lives in, the identities of their parents and close friends, where they spend their holidays and much more, was widely available to over 200 million people."[*]

Lady Sawers's Facebook page included all types of family pictures and information that linked the couple to controversial political figures and famous actors. The *Daily Mail* wrote, "Over the past year, Lady Sawers has been regularly updating anyone who cared to read her page—which could be found via Internet search engines—on everything from family parties and holidays to the health of their pets and her views on the crisis in the Congo."[†] Once discovered by the press, and questioned for the poor judgment of posting family pictures and family locations for the nation's top spymaster, the Sawers family removed the Facebook page and have cleansed their information from social media sites.

You probably aren't a spy, but you should still take steps to keep certain information about your private life private.

---

[*] Lucian Constantin, "MI6 Future Chief's Personal Life Exposed by Wife on Facebook," Softpedia.com, July 6, 2009.
[†] Jason Lewis, "MI6 Chief Blows His Cover as Wife's Facebook Account Reveals Family Holidays, Showbiz Friends and Links to David Irving," *Daily Mail*, July 5, 2009.

they trust should have access to our lives, relationships, and information. Luckily, there are several things you can do to protect your privacy.

Think before You Post

The first step toward greater privacy is to filter the information you expose about yourself. The less you say about your life, your family, and your activities in a public forum, the more privacy you will have.

Although this seems like a commonsense strategy, it requires some discipline to shift our thinking from posting anything we like to being more selective. Many people do not stop to think about the potential costs of stating their political or religious beliefs for everyone to see. We all feel our beliefs are sound, and we may not consider how offending someone who holds different beliefs may harm our chances for a new job or damage us in our personal lives. A potential new employer may feel that oversharing shows poor judgment. We may encourage diversity of opinion and lifestyle in our clubs, schools, and workplaces, but flaunting your beliefs and opinions can be counterproductive when those online statements are one of the few things about us that our new boss or teacher can discover.

Similarly, most people don't see the harm in posting pictures of how they spend their leisure time—for example, on a beach vacation or at a party, exposing more of their bodies and their lifestyles than other people might be comfortable seeing. If you wouldn't wear a Speedo swimsuit or a bikini to the office, why would you want all of your coworkers to see you in such a state of undress with the click of a mouse? Drinking a glass of wine with dinner is acceptable in our culture, but if all of your social media pictures involve alcohol, then someone just learning about you might believe that alcohol is a driving force in your life. Cultivating a professional image takes care and attention. An unattended online persona can undercut all the work you invested to appear professional to your colleagues.

New tools such as Foursquare, Yik Yak, and Shout allow you to share your physical location so that anyone can see how much time you spend at bars or casinos, with your boyfriend, camping with friends, waiting in line at a club, or far away from home. You may believe that sharing such information with the important people in your life can make your existence safer, more connected, and more interesting. However, sharing these data with everyone may be irresponsible, not only in encouraging crimes against you but also in leaking details about your private life that could affect your career or your future relationships.

Just because you *can* post information does not mean that you *should* post it. Posting online is an act of publication, similar to submitting a press release or picture to your local paper. Before you release a picture, opinion, location, or other information, you should ask yourself who might be able to see it and what that person might do with the information.

Imagine Your Audience . . . Smaller

A basic feature of all the major social media platforms is that you can control how many people see your information by adjusting the privacy settings on your page. Begin addressing your existing Internet persona by tightening the viewing circle on all of your social media sites. Whether you have an online diary, shared picture stream, or a standard identity page for business or pleasure, make active decisions about who can view your posted and uploaded material. Certain information should be for "friends only," although it's okay to leave other content open to the world. Try to make a conscious decision about which information is protected and which is made public.

As social networking sites mature, their administrative and privacy settings grow more sophisticated. At one point in the history of Facebook, a user only had the choice of making his or her entire profile public or making the entire profile viewable only by people marked as "friends." Now the administrators of Facebook allow a rich set of privacy options. Website owners provide these settings for your safety and convenience, so use them.

As of this writing, Facebook breaks your profile and onsite activities down into nine categories, including "your status, photos, and posts" and the geolocation application of "the places you check into." You can determine whether everyone who visits Facebook can see any of this information, only your friends can see it, or if it is available to a broader, but still limited, group called "friends of friends." This means that you can allow everyone to see your biography and favorite quotations, while keeping, for example, your religion and your direct contact information restricted to a closer circle of friends. To make these selections, go to the "Account" tab at the top right of your Facebook page, click the Privacy Settings tab, and customize your access settings.

Other social sites such as Google+, LinkedIn, and Tumblr also offer different privacy options that allow you to select which information is shared with various groups of site users. Some of these settings let premium paid users see more about you. LinkedIn, for example, allows paid users of its site to find out more information about all of its profiled users than you could see with a free profile and membership. Paid users get to see expanded profiles of everyone on LinkedIn and are allowed to "get the real story on anyone with Reference Search." Instagram offers a selection of frequently asked questions to help members decide how to use privacy features and to demonstrate ways of limiting access to a user's account.

Cleaning House

Do you hoard knickknacks and papers in your house? Maybe you're an online hoarder as well. You have to analyze what is already posted online and carefully clean out and remove all the unnecessary clutter.

One of the advantages of the Internet over older forms of media is that the Internet allows real-time input on any aspect of life. Many of our comments online are significant, if at all, only to the news of the moment, with no lasting relevance for our lives or the rest of the world. Telling your friend that her new haircut is lovely, telling your poker buddies that you will be late to the game this week, and telling the world that you are grieving for victims of a flood or earthquake—all of these statements quickly become clutter after their useful time has passed. You have no reason for keeping them as part of your online persona days, months, or years later. Dump them.

HIGHLIGHTED TOOL: X-PIRE!

Humiliating pictures seem to last forever on the Internet. Funny at first, the photo of you with the tomato sauce on your cheek or with the completely demented look on your face loses its appeal quickly but may remain for years online, haunting and marring your online persona. With this problem in mind, a group of German researchers created a software service called X-pire!

By using a free app download, a user of Facebook or Flickr can set an encrypted timer on his or her online picture that stops displaying the photograph after a predetermined expiration date. By using this tool, you can allow all your friends to see the photos of the bacchanalian New Year's Eve party but set those pictures to expire, dropping offline after two weeks so that you and other partygoers won't be embarrassed by them in the future.

The service creates sophisticated encryption technology to limit viewing of a photograph, and the X-pire! developers are working on making the service easy to operate for the average social media user. The founder of X-pire!, Michael Backes, has said, "The software is not designed for people who understand how to protect their data but rather for the huge mass of people who want to solve the problem at its core and not to have to think about it anymore."*

X-pire! is an app that works on both the Apple and Android mobile platforms. Also, although a picture marked to expire by X-pire! software cannot be viewed online after expiration, Internet users who know the picture's location could still download the picture and view it once it's saved on a hard drive. So this tool is not yet a panacea for people hoping to hide all their photographs over time.

However, the existence of X-pire! shows that researchers and companies are paying attention to the problem of long-term Internet exposure of personal photographs. Since the first publishing of this book, X-pire! broadened its abilities to automatically take down online information before it becomes stale or outlives its usefulness.

---

* "'Eraser' Software for Web Photos Launches," FoxNews.com, January 13, 2011.

Remember that information online is virtually permanent, and older information can give a stale and imprecise impression of your current life. Think of the Internet as a huge closet of data that you have to clean out now and then.

Take the time to analyze, delete, and archive older information, so that only the freshest data about you are available online. There are tools that automate this process, from Privacy Protector, which sweeps and manages your browser lists erasing your Internet browsing history, to X-pire!, a service that makes certain information and pictures fall off of the Web after a set period of time. You can moderate comments on your Flickr photo albums and videos with the Flickr Cleaner tool, and you can strip your online comments of links and HTML tags with the Comments Cleaner browser plug-in. TweetDelete allows a Twitter user to eliminate posts that are more than a month or two old. Just set the type of tweets to be deleted and the time frame for expiration, and you are ready to show the world a cleaner Twitter account. Learning how to use these tools can help you to keep your online persona clean and up-to-date without a significant time commitment.

Managing Friends

Remember, no man is an island; your friends and family are out there posting information about you, too. Try to clean your old information off friend's pages as often as you wipe it off your own. This isn't always an easy task because you first have to get your friends to understand all that you've learned from this book and why you and they should be concerned. Your friends and family may resist your requests or may not get around to taking down information immediately, so be relentless and educate them about the importance of a clean Internet persona. Tell your friends that you are undertaking a systematic clean-up of your online image and that you will help them scrub content off your pages when they decide to police their own images.

First, examine all of the pictures tagging you, the comments discussing your great test score or insight in the book club, and the shout-outs to you on the sites of others. Next, decide which items are worth the hassle of a take-down request. Prioritize pictures or comments that are particularly embarrassing or that will be noted negatively by your future mother-in-law or by the human resources director when you are looking for a new job. Think about how you want your online persona to appear and request that friends and family members remove those items that don't match the image you want to portray.

Separate Personas

Some people create alter egos that they wish to keep entirely separate from their business persona or the identity that they show their family. One thrill of the Internet is your ability to participate as yourself, or as VikinginDuluth, or as a third-level magic elf in a multiplayer role-playing game. You can keep a personal account for your friends and a business account for your clients. By taking reasonable care to keep them separated, your business contacts will likely not be able to track you to your friends, or vice versa.

It is important to remember that if you are one of the people who wants to keep online identities separate, you should create a new e-mail address for all sites and accounts for which you want to maintain anonymity. Using a single e-mail address to support your business at BobtheAccountant and your secret account at studmuffin4U is a dangerous proposition. Many Web investigation tools, including the social media search tool Spokeo (see chapter 4), can trace dozens of accounts that are all tied to the same e-mail address.

Because Hotmail, Gmail, Yahoo!, and others will provide you with one or more e-mail addresses for free, it's not expensive to ensure that each persona that you want to keep separate from the others is registered using a unique e-mail address; do not cross-list the addresses for online accounts. Keep Julie Lee separated from luvgrl93 before your worlds collide, to the embarrassment of everybody.

Removing Information from Third-Party Sites

Some of your worst exposures could be on the sites of the business you work for (think of last year's holiday party) or the organizations that you belong to (remember the summer pool party that your neighborhood association threw this summer?). Organizations are usually responsive to take-down requests, but these often must go through a vetting process and often take time.

You could also have been caught by a camera by a news reporter or party site. These pictures or posts are much harder to get removed.

Certain online data are under your direct control. However, much of the information that encompasses your Internet persona is within the control of companies, governments, and other people. In the previous section of this chapter, we discussed how you deal with asking friends and family to remove information from their sites; now it's time to examine how to approach companies or people that you don't know and ask them to delete your data.

Certain information will not be removed no matter how forcefully you demand it. Your county keeps records on real estate transfers and records naming the people who spoke at commission meetings, for example. It's not within the discretion of the county's webmaster to remove this information. Similarly, public corporations have filing requirements that may contain information about your stock ownership or other formal relationships with the company, and the Securities and Exchange Commission may also post stock purchase records on its website. These entities will continue to display those documents as long as you continue the relationship.

If you've been mentioned in a news story, the article or video will probably be placed in searchable archives. Even if you prefer that the world forget your first three marriages, the wedding announcements in those archives could keep them alive forever on the newspaper's website.

Other sites make a living off posted content, so they would not be likely to drop a picture from the site based on your request. If you find a risqué picture of your daughter in a swimsuit or a video of you entering an adult bookstore, a number of Internet sites would never be responsive to requests for removal because their revenue is based on finding and publishing photos and videos that titillate people or tattle on them. In some cases, the sites are poorly managed or simply unresponsive. With no incentive to take the time for removing the content you don't like, they will not make the effort, and you are unlikely to have leverage to force them to do so.

With limited exceptions, U.S. law does not provide a remedy for you to force a website owner to remove a picture of you from its site, as long as that picture was accurate and taken in public, no matter how embarrassing the picture may be. Many sites will honor your request to remove content featuring you, and making yourself a nuisance can often sway the webmaster's thinking, but threatening lawsuits may be counterproductive. You are more likely to reach the desired result with sugar rather than vinegar. Threats are little more than annoying and ineffective when you do not have the legal support to carry them out.

With regard to removal of comments that you posted on someone else's website, a U.S. court is likely to hold that you donated your comments to the site and the website's owner may treat them in any way it pleases. However, if you are concerned about such content, you should carefully read the online "Terms of Use" for the website in question. Frequently these terms discuss how your information will be treated and may provide an address to direct concerns when you want something withdrawn from their site.

You are likely to find that under the website's Terms of Use, all information posted on the site by any person belongs to the website operator and that operator has reserved the right to treat such information in any way he or she desires—keeping it displayed, removing it, or using it in a different context. The Terms of Use are not definitive statements of law, but they are often cited by courts as the only written contract between the site operator and those people who interact with the website. You should review the Terms of Use for sites where you deposit comments or content so that you have some understanding of how the site's operators will treat that content if you want it removed.

Just because a website provides itself with broad rights over your submissions does not mean that it will ignore your pleas to remove the content. For example, CBS Sports, one of the most popular sports fan sites on the Web, reserves to itself an almost comical amount of power over user submissions to its sports blogs and comment pages. The CBS Interactive Terms of Use states,

> When you provide User Submissions, you grant to CBS Interactive and its affiliates and partners a non-exclusive, worldwide, royalty-free, perpetual, irrevocable, fully sublicenseable license to use, reproduce, archive, edit, translate, create derivative works of, make available, distribute, sell, display, perform, transmit, broadcast and in any other way exploit those User Submissions, and any names, voices, likenesses and other identifying information of persons that is part of those User Submissions, in any form, media, software, or technology of any kind now known or developed in the future, including, without limitation, for developing, manufacturing, and marketing products. You hereby waive any moral rights you may have in your User Submissions.[1]

CBS provides several paragraphs of detail about how they plan to treat your submissions but also agrees to allow you to remove them when you choose. In its Terms of Use, CBS likewise provides contact links for you to make complaints or requests relating to the website.

Most responsible webmasters and most well-regarded commercial Internet businesses will respond to content removal requests. One of the most important legal protections for the operators of Internet media sites is the safe harbor provided by the Digital Millennium Copyright Act. This act requires that sites remove content that appears to violate a copyright. The act also provides legal protection for a website operator who acts promptly to remove certain content once it receives notice that the content infringes on a copyright. Courts have upheld this law for treatment of online subject matter that may violate rights other than copyright, including defamatory content, offensive content, and content that violates laws (such as child pornography). For this reason, most responsible media sites have instituted a procedure for taking down material when a person makes that request.

Once you formally request that a comment about you be removed because it is defamatory or it violates another legal right, an enormous company such as Google or Yahoo! would first send a message to the person whose comment you are asking to remove, detailing your accusations and asking for a response. If that person does not respond or is unable to adequately defend the content he posted, then the website operator will likely remove the content from its site.

Google was the target of the Italian government, which accused the company of allowing illegal content to be uploaded and to remain on the Google site. The criminal suit called into question the standard method that Google and many other sites use to decide issues of content removal.

Someone uploaded a video onto the Google Video site that showed several boys tormenting a child with Down syndrome. When requested to remove the video from its service, Google did so. But that wasn't good enough for Italian authorities, who pursued a criminal action against Google and several executives of the company on charges of violating personal privacy because they allowed the video to be posted and didn't remove it quickly enough. On February 24, 2010, a court in Milan convicted three Google executives and imposed a suspended six months' jail sentence on the Google decision makers. More than two years later, an appeals court in Milan overturned the judgment and vacated the Google executives' jail sentence. Knowing that a U.S. safe harbor exists for taking down offending content when requested, and also knowing that courts can impose possible penalties such as six months spent in an Italian jail for delayed removal of offensive content, many commercial and media sites now respond more quickly to such requests.

Simply asking a site to remove pictures, comments, or video can frequently bring the desired response from website operators. Often their self-interest dictates that they should pull a picture down rather than keep it online at the risk of a lawsuit. Of course, it helps if you can show a legal reason for wanting the content removed. If you claim that the content defames you or breaches your intellectual property rights, then you are more likely to get action. If the content is merely embarrassing, then the site displaying it will be less likely to remove it. Many sites are much more likely to allow easy removal of your own comments and content, but a request to pull down the content provided by other people triggers a longer and less accommodating process.

If an e-mail or phone call or letter to the people who run the website doesn't work, you may want to take the step of asking a lawyer to send the request on your behalf. For many companies, receiving a lawyer's letter places the request in a different category because it increases the possibility of litigation. To many executives, the involvement of an attorney in a dispute raises the stakes of ignoring the request. Also, the more powerful and conservative legal department handles letters from attorneys, rather than the customer service organization, which often has little power to make a change. Therefore, many online companies that ignore your personal plea to remove content from their sites will respect the same request from your lawyer. Check to see if you qualify for the right to be forgotten under the laws of California or the European Union as discussed in chapter 8.

Bolstering Your Image

If you want to look sharp, you pay attention to your wardrobe, eliminating the old, tattered, out-of-fashion items and adding new, up-to-date attire. Similarly, you can keep your Internet persona looking sharp by cleaning out the old irrelevant content, and the content that may describe your former self—before the job, before the move, before the wedding. Next, it's time to add fresh, positive content. You can take a page from the corporate playbook to regularly add affirmative content to the Web that can shape the way people think about you. If the most recent items about you are five years old, viewers will justifiably wonder what you have been doing in the meantime. Finally, make sure people can easily find that positive content.

When you're done you can step back and admire your work. You've never looked better.

You can follow certain models for creating new, positive content about you online and building the persona you want. You can also make sure people can find that content easily.

Do What Corporations Do

For more than a decade corporations have worked hard to manage their images on the Internet. They hire people to read blogs and complaint sites, to analyze news stories, and to lurk on hacker message boards to find negative posts about the company. They buy software and services that tell them every time they are mentioned and in what context people are chatting about them. And when they perceive a weakness in perception of the corporation or a problem that won't go away, companies take proactive steps to manage their online reputations.

Companies encourage their executives or their customer service professionals to create blogs about company products. They develop Facebook networks and show outtakes from their television commercial campaigns on YouTube. Corporations reach out to customers with online sweepstakes and enter-your-own-song contests. They hire other companies to manage emergency messaging or search engine optimization so that customers see the best possible corporate face.

Many of the corporate tools, tricks, and transformations are not practical for regular people to use in managing their online persona. For example, it is unlikely that you will be establishing a corporate-sponsored online chat forum or hiring bloggers to speak highly of you. In addition, you probably will not be proactively influencing an industry segment by carpet bombing online sites with positive news about you. However, you can learn about managing your online image by observing how corporations manage theirs.

For example, some of the most well-regarded consumer companies operating in the United States will take a proactive approach toward specific people who speak negatively about them online. They know that some critics can never be appeased, but others simply need more information and attention; those critics need to see that a targeted company cares about their concerns. So the targeted businesses directly address what they believe to be unfair criticism by engaging with the critics and finding out what can be done to make those critics feel better about the company. Often such criticism is a cry in the wilderness for a customer who feels ignored, and in those cases, direct and honest contact can help remove unwanted bad publicity from the Web.

Similarly, when you find information trashing you online, stop and think about the motivations of the writer. Although you will never be able to influence some people, many critics will appreciate a direct and honest approach requesting that the offending criticism be removed from the Web. The Internet's impersonality often makes it easier for people to comment in ways that they would never use in person. Confronting those people in a polite and honest way may be the best method to win their approval and loyalty.

A related corporate strategy involves joining an online conversation that is likely to affect you. If you, your family, your business, your military unit, your church or temple, or your neighborhood is discussed in any detail in an online forum, chat room, or discussion thread, you could join the conversation and steer the discussion in the direction you think it should follow. Actively participating in a discussion could be the best way to manage your online reputation.

Create Positive Content

One of the most common, but least discussed, corporate strategies for online image-building involves creation and Internet publication of new, favorable material so that the positive information offsets any negative information, or the positive information pushes the negative information so far down on the natural search results for Google, Bing, Yahoo!, and other search engines that people hardly ever see it.

Pushing a troublesome story or picture about you down one page has a stupendously significant implication on the number of people who might see it. Pushing the same item to the third page nearly drops it entirely off the map. A company called Chitika Insights studied click-through rates of each position on the first and second pages of a search engine query. They found that 94 percent of users clicked on an item that appeared on the first page of search results, and that less than 6 percent of people clicked to the second page of results and selected an item on that page.

Knowing this fact, you can see how adding a few items of relevant content about yourself to the general Internet pool of data can change people's opinion of you quickly.

Creating your own data is easy enough. Simply develop social media profiles and information on websites that discuss the aspect of your life or your online image that you want to address. The more information you add in different locations, the greater the chance is that people looking for data about you will see this information.

When asked how a person can communicate a positive image online, Baltimore-based corporate image consultant Henry Fawell advises:

> Inoculate, inoculate, inoculate. Build resistance to harmful content by being proactive, using good judgment, and using what is essentially free advertising in today's technology. In addition to building profiles at well-known social media outlets, identify blogs and media sites that cover areas of interest in your career or life and publish thoughtful comments on them. For instance, I will occasionally comment on articles at *Harvard Business Review*'s blog, charitable sites, or other strategic communications blogs. Doing so aligns me with positive brands and causes. I will also include a link to my company's website to generate free traffic.

The more positive material exists about you on the Web, the more likely someone will develop a positive impression when they look for information about you.

Positive information can be as simple as a description of your specialty at work and how much you love growing better at your job. It can mean a full discussion of a charity you support or one of your hobbies. Posting positive data about yourself may be as easy as providing a list of your schools and affiliations, or any awards you might have received. Show aspects of your personality that you want people to know about.

Help People Find You

Search engine optimization has been an obsession of many businesses since search engines became the primary means of finding information online. Some companies have taken extreme steps to move themselves up in search engine rankings and to make sure that a Google search quickly finds the information that the companies want you to see.

In February 2011, the *New York Times* published a story about the actions that retailer JCPenney took to manipulate the findings of Google searches. JCPenney admitted to using questionable techniques to fool the Google search algorithm, and Google subsequently punished JCPenney by dropping its links much further down on the natural search lists. The article also noted that Google had given BMW the "death penalty" in 2006, dropping its site completely from many search results because of "black hat" manipulation of searches that Google views as cheating.

Many of the companies you work with and respect also take steps to massage search engine results without resorting to the Internet equivalent of dirty tricks. You can learn from their examples, pushing negative information about you down onto the second, third, or tenth page of search results where it will be unlikely to be seen by anyone searching using your name.

Write!

Writing and posting new, positive content is important. You can open profiles on different types of social networking sites, including LinkedIn for business connections. You can publish information that you want people to see about you on your own Web page. Writing about general topics in your area of expertise can advance the conversation for everyone, but it is also a good way to generate the links that today's search engines prize. Interesting content brings

---

FEATURED WEBSITE: REPUTATION.COM

If you are deeply concerned about policing your online persona, then you can take a page from the corporate playbook and hire a specialist. Reputation. com, once known as Reputation Defender, offers services to protect your image online and to hunt down and remove problem data as they arise on the Web.

Founder and current chief executive officer Michael Fertik says he started the company because he was disturbed to see young people haunted by lapses of judgment, and he wanted to provide a service to help clean up an online reputation. He is concerned about the inaccurate impressions that anyone can receive online. Fertik was quoted in the *Washington Post* as saying, "Google's not in business to give you the truth, it's in business to give you what they think is relevant."*

Reputation.com offers a number of services, including MyPrivacy, which monitors the Internet for instances of your personal information published on different sites. The MyPrivacy service may request that the website operator remove the personal data in certain circumstances. Another interesting service is MyReputation, which purports to control what other people see when they search for you and helps you create and publish positive content to manage your online image. The company also has business products to assist companies in managing their reputations.

If the prospect of managing your online persona is overwhelming to you, a company like Reputation.com can help make the prospect easier for a fee. It might be the guidance you need to stay ahead of the Internet information machine.

---

* Susan Kinzie and Ellen Nakashima, "Calling in Pros to Refine Your Google Image," *Washington Post*, July 2, 2007.

in viewers, links, and higher search rankings. Use keywords that are likely to be used in searches in headings on your site or page and generate as much content as you can so search results are filled with positive images of you. One of the great aspects of the Internet is that it seems to value short, pithy, interesting content to long, boring detailed articles. This should make your writing job easier knowing you can pack a greater punch with a series of smaller articles and interlinking them.

Taking charge of your online persona and actively managing your image can be an easy way to present yourself in the best possible light online.

CHAPTER 6

PROTECTING IDENTITY IN A CRISIS
Identity Theft and Defamation

B ecause of certain characteristics of the Internet—the ability to perform tasks anonymously from a distance and quick, easy publication to the world—the number and variety of serious identity and reputation crises are growing, and you could be the next victim.

The Internet, like any area populated by humans, contains predators, pranksters, liars, and thieves. Although there are dozens of reasons why someone might want to adopt your Internet persona, in this chapter we will discuss the two most prevalent: identity theft, where another person pretends to be you to get money; and image impersonation, where another person pretends to be you for fun, malice, or to damage your reputation.

Identity Theft

Identity theft will continue to evolve and grow as a cybercrime of choice for fraudsters. One source of information that helps cybercriminals to steal your identity is a data breach. When they break into the data stores and treasure troves of information about you, they have what they need to unlock the keys to your identity.

The Size of the Problem

Just how big is the data breach problem and its downstream impacts on identity theft? Data breaches were up 46 percent from 2013 and hit a whopping worldwide total of 1,540 reported breaches in 2014.[1] That volume of breaches gets put in perspective when you realize that the data compromise involved more than one billion records.[2]

It is estimated that 54 percent of the global data breaches had an identity theft-based motive.[3] In fact, identity theft is considered the fastest-growing crime in the world. The worst ten countries for identity theft, from best to worst are: Mexico, the United States, India, the United Arab Emirates, China, United Kingdom, Brazil, Australia, Singapore, and South Africa.[4]

According to the 2015 Javelin Strategy and Research report, someone was a victim of identity theft in the United States every two seconds last year.[5]

The U.S. Federal Trade Commission (FTC) estimates that more than 330,000 identity theft complaints were received from consumers in 2015, making it the top complaint for the fifteenth year in a row.[6] The problem is so great that President Barack Obama has appointed an identity fraud task force.

Your identity is more vulnerable on the Internet than almost anywhere else because you regularly give your valuable information to people you cannot see and bad guys are continuously pinging your Internet connection, looking for weaknesses. If you have good credit, money in the bank, health insurance, citizenship, a passport, or own a house, someone would like to use your good fortune for their own purposes.

Crooks don't need to take your entire life to gain access to your credit or your health plan, they just need to collect a few important numbers and facts and then use that information to apply for new accounts or to drain yours. Certain sites charge criminals a flat fee to buy useful sets of financial or healthcare data that have been stolen or exposed, and anyone who buys those data is likely to use it for illicit purposes.

The Growth of ID Theft Online

We hit a major milestone in 2009. According to a Symantec press release, Internet crime outpaced illegal drug trafficking and sales as the major criminal profit maker. Symantec indicated that in 2009 a crime occurred in New York City every three and half minutes, whereas an identity is stolen online every three seconds.[7]

THE BILLBOARD RULE

This simple rule, also known as the "mom and grandmom rule," requires no software. If you would not want people to drive by your billboard and see something posted there (or you wouldn't want your mom or grandmom to see it), don't put it online.

Also, be guarded about how much information you post about yourself online, including answering those quizzes on Facebook that look harmless. Those posts and quizzes hand out information to friend and foe alike.

The Internet is global, and so criminals have built far-reaching global enterprises. According to the FBI's Internet Crime Complaint Center (IC3), one place where criminals watch for victims is on Internet auction sites such as Craigslist or eBay. The seller of an auction item sets up shop and appears to be U.S. based. If the only form of payment accepted is wire transfer to a bank, services such as Western Union, or an escrow service of their choosing, be suspicious. Such criminals often route dollars through Latvia, Belarus, or Romania.

PayPal released a study of identity theft online and found that citizens in Canada, the United Kingdom, and the United States are the most frequent victims of Internet identity theft. It's believed that this is because the United States, the United Kingdom, and Canada have the highest volume of online e-commerce.[8]

Your information, credit card number, and bank account data are bought and sold online as if they were the hottest item on the Amazon.com holiday wish list. The person running this criminal auction may appear to be the boy next door. One alleged criminal auctioneer was arrested in the United Kingdom. He was thirty-three years old, worked at a Pizza Hut, and hung out in Internet cafes. He created a site called "DarkMarket," and it allowed criminals to buy, sell, and trade private and personal information. When it was operational, the information they had for sale was stunning and included information that could allow a criminal to answer your bank account's secret password questions. Your identity is priceless to you, but for roughly thirty British sterling, information for all your credit cards may be available at a bundled discount.[9]

The site also offered its approximately 2,500 global customers online training to help them be better Internet criminals. Fortunately, the United States and the United Kingdom worked together on the case and were able to successfully close down the site, arrest the founder, and arrest members in the United States, United Kingdom, Russia, Israel, Turkey, Germany, and France. Unfortunately, replacement sites are popping up every day and are thriving.

Do you wonder how you can manage your personal risk in such a world? Take the privacy quiz at http://newsquiz.sciencemag.org/privacy and if you are in Canada and want to know your privacy rights, check out this privacy quiz at https://www.priv.gc.ca/youth-jeunes/quiz/index_e.asp.

How e-Commerce Sites Protect You

When you agree to accept a check from your neighbor for a basket of vegetables, you trust that the check will be honored because you know your neighbor. You recognize her by sight, you know where to find her, and you have a personal relationship. You also know that she would be embarrassed to bounce a check because that could have an impact on her reputation in your community.

None of these assurances exist when you spend money online.

If you pay for a jacket online, the e-commerce site doesn't automatically trust you. The site will only accept payment by credit card or a verified payment system such as PayPal because these methods are intended to minimize fraud and bad payments by performing an authorization check before processing the payment. If the credit card system authorizes a transaction, then the e-commerce merchant knows that this particular card is active, the person has sufficient credit available, and the card has not been flagged as stolen.

Sophisticated card issuers will even run the authorization request against algorithms that flag unlikely transactions and refuse authorization until they can contact the cardholder. If you have only made purchases in and around St. Louis, Missouri, for the past three years, when new purchases spring up in Mexico or Brazil, your bank will probably notice the unusual behavior and refuse foreign purchases until they can contact you. The card system itself serves as a method of minimizing fraud.

Where ID Theft Comes In

If a person directs his browser to an online store and he has your credit card number, the card's expiration date, and the security number listed on the back of your credit card, he can pretend to be you and make a purchase that is sent to *his* house, while the purchase is charged to you.

To make the fraud work, he might need your address, but he can find that in a public directory. The e-commerce merchant won't know that this person is not you. How could he? All a site knows about you is the information that is entered about you.

Even if you are careful about exposing your credit card, other people still have access to a great deal of information that allows them to pretend to be you. Many of the workers at your bank have this type of information. Workers at stores where you shop or waiters at restaurants where you eat have access to the data, and so do third-party card processors. Anywhere you use a card, its valuable information is exposed again and again.

If your identity is stolen online, the damage does not always remain online. An identity thief who has enough information about you can open new credit accounts in your name, establish phone service in your name, or file fraudulent tax returns in your name. She may even file bankruptcy under your name to avoid eviction from an apartment that he or she rented using your name. Some identity thieves steal information to qualify for healthcare payments or to get an identity that will allow an illegal alien to stay in the United States.

Protecting Yourself against ID Theft

Although credit cards provide a tool for thieves to steal your identity, your liability for any credit card fraud is limited. A more significant identity theft issue involves your bank accounts. Unfortunately, your bank may not limit your liability as credit card companies do. You just might find yourself fighting to prove that you were not the one who drained the money out of all your accounts.

The famous gentleman thief of the 1930s, Willie Sutton, claimed to rob banks "because that's where the money is." Whereas Willie Sutton was known to visit the bank manager's house at night and then walk in with him in the morning to open the vaults before anyone arrived, a modern-day bank robber would only need to steal your account number and then point his or her Internet browser to your bank's website.

Get in the Privacy Habit

You can do everything right and still be a victim of identity theft as a result of data breaches or other mistakes that are no fault of your own. Still, you should be informed about risks and what you can do to minimize them.

There are technology tools a crook can use to steal your information, however, it could be your own behavior putting you at risk. Be careful with your account information. Not everybody can remember all the account passwords and codes he or she needs for each online financial account, so some people write them down. It is better not to write down your banking codes, but if you have to, simply keep these numbers locked up and away from your computer where someone else can't easily find and use the information. You might want to write this information in the middle of a string of nonsense words and characters, so you only need to memorize the first and last letter in the chain.

Account access can be compromised by simple codes and good guessers, so get in the habit of choosing passwords that are long, and contain a combination of uppercase and lowercase letters, punctuation, and numbers. These would be difficult for someone to guess.

TAKING STEPS TO STOP THE SNOOPS

Your best bet for heading off identity thieves and would-be defamers is to stop people from getting to you or your information online in the first place. There are many ways to protect yourself from prying eyes. Here are a few easy and free ways that you can use to protect yourself from snoops.

Turn Off Geocodes. Every camera and smartphone that supports the feature provides options to turn off geocodes. Check your operating manual for

(continued)

instructions. A warning, though: If you turn off the geocoding function, you lose your ability to use the GPS function until you enable geocoding again.

Check Your Browser Settings. Depending on which browser you use, there are a variety of settings available to protect your privacy. Most browsers, with only a few mouse clicks, allow you to "empty cache," set security to "high," warn you before a cookie is installed, and set up "private browsing."

Anonymizers. An anonymizer, quite logically, helps to make you anonymous online. However, it does not work like an invisible cloak to turn you into an Internet user who leaves no traces. What it can do is help make your Internet surfing activity harder to trace. Anonymizers hide your surfing behind a proxy server, essentially another computer that acts as the go-between from your computer to the Internet. One example is Google Chrome. Google Chrome offers a feature called "Incognito Mode." Any cookies or tracking that is done while you surf the Net are deleted after you terminate your Internet session. Such a program can be useful in protecting your privacy and personal information. Unfortunately, it can also help evil-minded peekers and gawkers to hide themselves and make them harder to track down. And remember, these guys are experts at using this type of technology.

Do Not Track. The Federal Trade Commission in the United States has been considering implementing a "do not track" law for companies with a Web presence. The agency sees this law as a simple tool that would work a lot like the "do not call" phone registry for telemarketers. A lot of issues will have to be debated before legislation is approved. In the meantime, you can take matters into your own hands by using your browser settings. Many browsers currently either have a "do not track" option or they are developing one based upon customer feedback. One example is Mozilla's Firefox browser versions in 2016, which allow you to decide whether you want to be tracked as you surf the Net or conduct business online and includes a "forget button" removing browsing information from your history.

Technology Solutions. If you only have fifteen minutes, at a minimum try these two technology options:
- Browser settings: Decide how much privacy you need, make the appropriate settings, and test and check these settings regularly.
- Privacy settings on social networks: These are not foolproof, and technology glitches and new releases tend to reset your privacy settings so your information is available to the widest audience. Still, it's worth the time it takes to check your settings and make them as private as possible.

Password Overload

The most traditionally overused passwords around the world are still pretty common. *Password* and *123456* might be easy to remember but you are making it easy for hackers to get into your account if you use them.

SplashData released its 2015 annual "Worst Passwords List," and in the top twenty-five there are sports-themed passwords such as *baseball* and *football* and passwords such as *dragon* related to favorite TV shows like Game of Thrones.

The study reviewed all the leaked IDs and passwords published online, looking at more than three million passwords. Here are some key findings:

- Most of the hacked and popular passwords were six characters long or less.
- Many of the passwords were just a word that can be found in a dictionary.
- Passwords such as *superman*, *batman*, and *dragon* were new on the top twenty-five list this year.

Where do you keep track of your passwords? In a notebook? Maybe you have them written on sticky notes posted around your computer desk. Many of us choose that route because we aren't so sure what else to do. There are some new advances in storing information online that might make online vaults a better option for your passwords.

Online vaults store passwords for you and many of them will check sites to see if they are phishing or fake sites. Many vault services will also provide password generation to help you create strong passwords. Several services now require two-factor authentication (e.g., sending a text of a verification code to your cell phone on file) before you can open the vault.

Before you use one of these free or inexpensive services:

- Do your homework.
- Read the privacy policy.
- Type into your favorite search engine the name of the service and the word *breach* to see if any breaches pop up.

A few password vault services that you might try are:

- LastPass
- KeePass
- 1Password

Remember, all technology, including that used in password vaults, is hackable. To protect your information, follow these steps:

1. Protect your password manager with a strong password. It can be something that you write down and keep in your wallet.
2. Look for a password vault that times out and requires you to login again after a few minutes of inactivity.
3. Use two-factor authentication every time it is offered at a site.

Get Technology on Your Side

Be careful about letting others use the computer that you use to access your bank accounts, either in person or remotely. Someone who uses or has access to your computer via your Internet connection can download malicious software onto the machine and can use that software to capture your bank passcodes. In addition, it is possible to download keystroke monitoring software to your computer that allows a crook to track every password and account number as you enter them into websites. To stop the download of such programs, turn on a firewall, such as the one built into Windows and Apple operating systems.

Keystroke logging software is one type of malicious software, but it's not the only one. Malicious software includes computer viruses and spyware, which can also be downloaded onto your system when you open an e-mail attachment or click a link on a website. Follow this advice to get technology on your side:

- Be sure that you know the source of each file that you load onto your computer and that you trust the business or person that sponsors or offers the download.
- Install antivirus and antispyware software such as Sophos or Symantec Antivirus and be sure to keep the definitions for current malware updated on a regular basis.
- Your browser also offers protections if you set it up to block you from all but trusted sites and to flag sites that have a history of downloading spyware onto visitors' computers.
- Malware is often transmitted in attachments to e-mails that are sent to your computer. A virus can, in fact, be contained in a single pixel of a picture you click on in an e-mail. You should be cautious about clicking on links in e-mail messages that could take you to dangerous sites, and never open file attachments from people you do not know.
- Even some of your most trusted sites can have poison links, also known as *malvertising*. These are advertisements that lead your browser to dangerous sites. When you go to those sites, poison link developers try to download malicious software onto your system. Top websites such as the NewYorkTimes.com and Gizmodo.com (a technology site that boasts as many as three million page views a day)

have accidentally hosted advertisements that linked to malicious or fraudulent content.

- Some identity thieves have become sophisticated at developing software that takes over your account and sends out e-mails under your name with malicious attachments. Be aware that opening attachments from a friend's e-mail or a business associate's e-mail could also be risky, so if you're not expecting the attachment, you may want to call before you click.

- Remember that your browser or e-mail program can be set to block the opening of e-mail attachments or to look for suspicious junk mail.

Going Phishing

Phishing is a fraudulent activity that involves a criminal sending out e-mail that appears to be sponsored by someone familiar to you, such as your bank, insurance company, or an online retailer. The fraudulent e-mail will ask you to click on a link to access or update your account, visit a fraudulent site, download certain software, or enter your account number, passwords, or Social Security number. If you take any of these actions, you have just given the thieves what they need to steal from your accounts. Around the globe in the second half of 2014 there were at least 123,972 unique phishing attacks where the cybercriminals posed as a trusted person or organization.[10]

It's hard for the companies you know and trust to shut down phishing. Often criminals pretend to be someone or be employed by a company to lull you into a false sense of security. You can usually spot a phishing scam by some telltale signs, including the following:

- One or more links in the e-mail that you are instructed to click. Note that on a Windows computer, you can right-click on such a link and see its properties; does the address look like a legitimate business website?
- Bad grammar, spelling, or punctuation that an institution would never send out.
- An odd sender's address in the e-mail header.
- Missing or badly executed company logos.
- The e-mail is not addressed to you personally but to "Member" or "Customer."
- The message urges quick action involving some problem with your account, usually involving clicking on a link and entering your account information.

Never provide your most secret information—account numbers, Social Security numbers, passcodes, or your children's information—unless it's as part of a transaction on a trusted site. Trustworthy business contacts and merchants would never ask for this information if you are not completing a transaction on their sites, so call them and confirm that they sent the e-mail.

You can also go to their site to find out if a communication is legitimate, but don't do it by clicking a link in the e-mail; instead, type the URL for their website into your browser yourself. Be sure not to call the number offered by the e-mail for requesting the information. Find the phone number for customer service on your account statement instead.

If you do see a suspicious e-mail that purports to be from your favorite person, company, or organization, make an attempt to notify them. In addition, you can also do a global good deed by reporting them to the Anti Phishing Working Group (APWG). You can forward the phishing e-mail to reportphishing@apwg.org.

Sneaky Geocodes

Most of us know not to post pictures of the local sites while on vacation because that lets people know you're not home. You may know it's not a good idea to tag a lot of photos of your family at home with geocodes because people with ill intentions can track you down. But did you realize that you should be careful of the pictures you take of your pets, too? This applies to pictures taken with a smartphone or with a newer model point-and-shoot camera that embeds geocodes into the digital photos you take.

Owen Mundy, an assistant professor at Florida State University, created an app called "I Know Where Your Cat Lives." He randomly selects one million pictures from the Web that include the word *cat*. He then uses the picture's metadata to show the exact coordinates of where the cat was when the shot was taken. Try it out to see this important Internet identity lesson: http://iknowwhereyourcatlives .com.

I've Been Hacked!

Remember that your credit card information, Social Security number, or other personally identifiable information could be stolen when a company's own system is broken into. Since 2005, many well-known companies have reported that information was stolen by crooks attacking their systems (called *hacking*), by laptops being misplaced or stolen, or from theft by company insiders. The list includes TJX (the parent company of T.J. Maxx and Marshalls), MasterCard, Sony, Citibank, and the U.S. Veterans' Administration. The global breaches in 2015 that made the headlines with four million or more records compromised included: U.S.-based Anthem Health Insurance (78.8 million), U.S.-based Office of Personnel Management (21.5 million), U.K.-based cell phone company TalkTalk (4 million), followed by the Turkish Government (50 million),

Topface (20 million), Hello Kitty's user database (3 million), VTech toy maker's database (6.4 million), and Gaana.com (10 million).[11]

EXPERT ADVICE

Todd Inskeep is a cybersecurity expert at Booz Allen Hamilton and president of Incovate Solutions, LLC, in Charlotte, North Carolina. His previous jobs include data security positions for Bank of America and the U.S. National Security Agency. We asked Todd about how merchants and banks guard against identity theft, and Todd responded with this comment:

Merchants generally trust that if someone is ordering something they want to receive the product or services, so the address, e-mail address, and even credit card information is usually valid. However, to prevent and manage fraud [or identity theft], merchants also check this information against commercial information sources. For example, the U.S. Post Office and others sell information about valid home addresses that merchants can check—so they don't send boxes to non-existent homes. The credit card companies let merchants quickly check valid credit card numbers and the security code at a low cost, further encouraging the merchant. Some merchants track other information like the Internet address of your computer (IP address). They might suspect something if your home address is Detroit, but your computer is based in Budapest. Merchants also use the credit or debit card billing address and other information to try and validate that it's really you buying their goods.

Bigger merchants like Walmart can manage fraud better than a smaller company. And some small companies could literally go bankrupt from a single large fraud.

In many cases merchants will allow a purchase and ship the goods as the apparent purchaser instructs. In those cases the credit or debit card purchasing rules protect you, the consumer. Usually when you report any fraud, every bank and credit union I know of refunds you promptly, usually within 2–3 days. In rare cases, like if you waited six months or a year to report a fraud, the bank might limit the refund. Then the bank and the merchant determine who actually loses money based on the credit card operating rules.

Online banking is actually one of the best ways to combat identity theft. It lets you check your money and history frequently so you find unexpected problems quickly. By checking frequently you can avoid spending too much. Most banks offer online bill payment, [which] is even better. You can lower the costs of writing checks, buying stamps and mailing bills. You also avoid mailing checks which can get lost or stolen, leading to identity theft. Check fraud is actually a much bigger problem than online credit card identity theft.

Data breaches at service providers such as the e-mail marketing firm Epsilon affect customers of many well-known companies. In these cases, the affected business is likely to notify you that your account was compromised, and your bank may issue you a new credit card or a new account number. However, there is no substitution for reviewing your account statements every month and for checking your credit reports at least once a year.

What to Do If You're a Victim

When it comes to credit card fraud, the system itself provides protections for you. A victim of identity theft should quickly order a copy of his or her credit report to check accounts for financial transactions that he or she doesn't remember. As soon as you notice a problem on your bill or that your card is missing, you can cancel all transactions, or only those that you don't recognize. Under the bank's credit card contract, you will only be liable for fifty dollars of fraudulent purchases made in your name as long as you carefully police your account. Faithfully read your statements each month to catch the fraud right away. If you alert your bank to any fraud in a timely manner, it's likely to credit all the fraudulent purchases back to your account, minus the fifty-dollar maximum charge. Although requesting a replacement card and updating the new card number for your various accounts is a hassle, it beats paying your life savings for someone else's purchases.

Beyond credit card fraud, you have certain legal protections. Identity theft itself is a crime. While you could file a lawsuit in civil court against your identity thief if you can find him or her, you should first treat any serious identity theft as a criminal matter. If you are a victim of identity theft, file a police report with your local authorities.

You should also follow the instructions and recommendations of the FTC site on identity theft (https://www.identitytheft.gov). The FTC not only provides the most up-to-date information on fighting identity theft and managing your life once your identity has been stolen but it also includes specific sites and addresses to help you. The FTC site discusses products and services that you might obtain to help clear your name, and it answers the most common questions that people ask when their identities have been stolen. The site includes information for businesses that have lost their customer's information and resources for law enforcement and anyone else who wants to fight identity theft. The FTC's guides to detecting, deterring, and defending against identity theft include videos and are easy to understand. In short, this site should be your first stop for protecting yourself in a suspected case of identity theft.

You should also immediately file a report with the FTC when you know that your credit and name are being used in an identity theft scam. Sharing your information with the FTC will help law enforcement track and capture the thieves. You may have been caught up in a large-scale fraud and, if so, the report you file with the FTC might help solve the crime or return your money to you.

When you are attempting to convince banks, retailers, or others fooled by a thief who abused your account, each merchant or bank will ask to see a copy of the police report and the FTC report that you filed concerning the identity theft. They ask for this because they know that it is a crime to file a false police report, so if you can show a copy of your report, then you are more likely to be telling the truth to them and not merely attempting to escape from paying a debt.

If your identity has been stolen, you should also immediately shut down all accounts that were opened in your name by the identity thief. The more quickly you close these accounts, the faster you will stop the identity thief's activities.

You should also place a fraud alert with all three of the major credit reporting agencies: TransUnion, Experian, and Equifax. This will notify all prospective creditors that someone is using your name and credit to commit fraud and will stop the thief from opening additional accounts in your name. An initial fraud alert usually lasts for ninety days, which should be long enough to allow you to clean up your credit.

You can also ask the credit reporting agencies to include an extended fraud alert in your file, which could be active for years. Fraud alerts should be offered to you for free, and each of the credit reporting agencies also offers additional credit protection services, usually for a one-time or monthly charge. Under a fraud alert, you will receive notice from each credit-reporting agency when any new accounts are set up under your name. The free alert is likely to be enough to protect your credit by stopping new accounts from being opened in your name, but it will not stop ongoing identity theft of your present accounts.

You can take the additional step of placing a credit freeze on your record. Under a credit freeze, no one can process a credit application in your name unless you lift the freeze. This action stops criminals from opening credit accounts in your name. You can temporarily lift a freeze at any time if you want to apply for a credit card, car loan, or mortgage, though this might cost you a small fee.

What Do Privacy Laws Protect?

The last twenty years have seen explosive growth in the laws designed to protect privacy of personal information. However, much of the information that you might consider private or sensitive is not guarded by the laws of the United States, and the fact that companies retain information about you or that people can find your data online is only legally actionable under limited circumstances.

AT LEAST YOU HAVE YOUR HEALTH

In the United States, many of your online activities are not protected by privacy laws. The law protects only certain types of information that is used in certain ways. For example, information regarding your health is protected when you provide it to medical professionals. However, that same medical information may not be protected if you share it online in discussions with Facebook friends or with a website that asks you to take a quiz about your physical fitness.

If you explain the state of your pregnancy to prospective employers at Monster.com or to a travel insurance company online, that health information may not be protected under law. Similarly, if you share your genetic information with an online company promising to provide information about your ancestry, the data are not protected by health information laws. The law only protects information you provide to certain healthcare professionals such as doctors, hospitals, and pharmacies so that they can provide healthcare analysis or treatment.

Keeping Finances Secure

For most people, the most important legal data protections for information they post online relates to their finances. When you shop online, you generally provide your credit card information. The store has to use the card information for a specific transaction, and nothing more, unless you give the merchant permission to keep the card on file for future purchases.

Banking information can only be shared in specific circumstances that are intended to benefit the banking customer. Everyone who would be exposed to your financial information from an online transaction—the merchant that takes your credit card, the banks that complete the money transfers, and the companies that operate the payment systems and/or that process the transaction behind the scenes—all are strictly regulated on how they can use and share your data.

These financial data laws have been tested in situations in which merchants did not intentionally sell or transfer customers' financial data but instead hackers broke into the business's computers and stole data.

If your financial information is exposed by or stolen from an online company, you will receive notice of the security breach and your bank will probably issue you a new credit card and cancel the compromised card.

CASE IN POINT

Between July 2005 and 2007, the website of retail company TJX, owner of the T.J. Maxx and Marshalls brands, was hacked by professional criminals. TJX admitted losing 45.7 million customer account records, including the payment card information from these customers, while banks claimed that more than 94 million customer credit cards were affected.[*] Anyone who shopped at T.J. Maxx, Marshalls, or one of the other TJX stores (either online or at the physical store locations) during this two-year period had their financial information exposed to criminal hackers. If your card data were lost, you might have been the victim of identity theft.

A total of forty-one state attorneys general sued TJX based on this breach and collected nearly ten million dollars in settlement, but only a small fraction was used to assist people whose information was lost.[†] One of the gang of hackers who committed the crime was caught, arrested, prosecuted, and sentenced to twenty years in prison.

---

[*] Dawn Kawamoto, "TJX Says 45.7 Million Customer Records Were Compromised," CNET News, March 9, 2007; Ross Kerber, "Banks Claim Credit Card Breach Affected 94 Million Accounts," *New York Times*, October 24, 2007.
[†] Robert Westervelt, "TJX to Pay $9.75 Million for Data Breach Investigations," SearchSecurity.com, June 24, 2009.

Although laws exist that can force banks, merchants, and processors to protect your payment and other financial data, those laws are relatively new. The results of suing under these laws are uncertain at best. People have sued merchants and banks for losing financial information but are often unable to prove direct damages to receive significant compensation for the loss. A customer whose financial data is exposed may be best served by closing the account and requesting that the negligent party who lost the data pay for at least a year of credit monitoring services to guard against identity theft arising from the loss of information.

Protecting Children

U.S. law protects children's information online and requires that parents be notified when children twelve or younger sign up for contests or accounts on websites that will contact the children by e-mail. You have the right to insist that a company not contact your child or that it send messages to your child through your e-mail account.

One of the practical problems with online child protection is that, even when a business is being careful and follows the law, kids often lie about their age to gain access to websites. In effect, the business has no way of knowing that its new customer is a child.

This is one of many reasons to watch your child's Internet usage carefully. If your child leaves an Internet trail that wrongly lists his or her age as older, his or her own Internet persona will be distorted and could expose him or her to adult risks that he or she is neither mentally nor emotionally prepared to handle.

Differences around the World

Many countries are much more protective of an individual's personal data than the United States. U.S. laws have grown to reflect a protection of business interests balanced with the interests of individuals to keep certain classes of data private. Other countries, such as Canada, Mexico, and the nations of the European Union, regard the privacy of sensitive data as a human right to be protected from business and government in nearly all instances. In these jurisdictions, a business that takes personal or sensitive information from a citizen can only use the information for the reason it was offered and must receive permission to do anything more with the data or to pass them on to third parties. As evidence of this, in 2016 the European Union approved new data protection regulations. The new laws state that if a company violates the consumer protections they put into place, they could be held liable and be forced to pay up to 4 percent of their revenues in reparations to the aggrieved consumers.[12]

Understanding the Responsibilities of Websites

You can help to protect your own information by understanding how the websites you visit intend to treat you and your data. Most commercial sites that you visit on the Internet have a posted privacy policy that explains how the site's owner uses the data you provide. These privacy policies are policed by state attorneys general and by federal agencies, so they must be accurate as a company can be subject to significant regulatory penalties if they aren't.

For example, in early 2011, the FTC signed a consent order with Google subjecting the company to independent privacy audits every two years for the next twenty years. The FTC claimed that a now-terminated service called Google Buzz treated information differently than was explained in its privacy policies. Commenting on the Google settlement following the FTC's investigation, FTC chairman Jon Leibowitz said, "When companies make privacy pledges, they need to honor them. This is a tough settlement that ensures that

Google will honor its commitments to consumers and build strong privacy protections into all of its operations."[13] The FTC has even filed claims in bankruptcy court to stop the sale of a bankrupt company's consumer information collected from its website because the sale would violate the defunct company's online privacy policy.

When you review an online privacy policy, look at what information the company will collect from you and what limits the company places on the use of this information. Many companies claim that they will never share your personally identifiable information with any other entity. This is the strongest protection that you can expect from any website operator. Other companies will claim to share your personal information and addresses only with "affiliated companies" or "marketing affiliates." You may want to contact the site to learn what they mean by these terms. Other sites will not limit their sharing of your personal information and may be selling your data to any buyer. You may decide that it would be wiser to refuse to set up an account or purchase goods from sites that are willing to share your data.

Some companies also discuss their data policies in their published Terms of Use. These documents will frequently describe the level of control that a commercial or government site keeps over information on that site and how it shares information with vendors, advertisers, and marketing affiliates. If you are concerned about how a website will use your information, you should always check the privacy policy and Terms of Use, where the site operator is likely to explain, or at least hint at, the rules it intends to follow.

Some companies protect your privacy better than others, but based on a study, "Ranking Digital Rights," no company is a true standout.[14] This study reviewed the privacy policies and user agreements of many big tech companies and graded them based on how well they protect you. The survey was run over approximately two years and reviewed end-user service agreements, privacy policies, terms of service, and corporate reports. The survey administrators asked thirty-one questions based on the data they collected. They then provided the findings to each company for input. Not one company in the report received a grade above a "D." The highest grade went to Google with 65%. However, the good news is that tech companies are improving in how they disclose information about you and how they might use your data in the future.

Image Impersonation

In the fall of 2010, as former congressman and White House Chief of Staff Rahm Emanuel was campaigning to become the first new Chicago mayor in more than twenty years, a new Twitter account appeared under the name @Mayoremanuel. Its profile picture showed Rahm Emanuel thumbing his nose, and the tweets, each containing constant and outrageous profanity, told a story that sometimes tracked the real Rahm Emanuel's daily routine and other times veered off into absurd adventures such as living in an igloo within Chicago city limits and cultivating celery plants with Mayor Daley to make celery salt in a greenhouse on the roof of the Chicago city hall.

The fake Twitter feed soon had many more Twitter followers than the real Rahm Emanuel's Twitter account. This imposter Twitter account was described by the *Atlantic Monthly* as "next-level digital political satire and caricature, but over the months the account ran, it became much more. By the end, the stream resembled an epic, allusive ode to the city of Chicago itself, yearning and lyrical."[15] The real Rahm Emanuel offered a $5,000 donation to charity if the author impersonating him on the @Mayoremanuel Twitter would reveal his identity.

Not every case of online impersonation is clearly meant as good-hearted satire. Someone pretending to be you online can insult your friends, accept invitations on your behalf, and make rude comments to members of the opposite sex—all in your name. In short, an online impersonator could ruin your reputation. If information about you is false and is harmful to your reputation, it may qualify as defamation.

The Face of Online Impersonation

Internet image impersonation is easy to do. Anyone can open a free e-mail account with Yahoo!, Hotmail, Google, or any other e-mail provider and use your name. Setting up a social media account on social networking sites such as Facebook or Myspace is equally simple. With a little information about your life, your impersonator could even fool those people closest to you.

Unfortunately, it can be difficult to remove these accounts from the Internet. Most online companies assume that an account is opened in good faith, and you will probably have to prove the damage was done by an imposter (and prove that the imposter is not simply another person who happens to have the same name) before a site such as Yahoo! or Facebook would consider closing an active account.

In some cases, the law provides extra ammunition against online impersonators. For example, under California law, it is now illegal to impersonate another person online. The statute states, "Any person who knowingly and without consent credibly impersonates another actual person through or on an Internet Web site or by other electronic means for purposes of harming, intimidating, threatening, or defrauding another person is guilty of a public offense."[16]

This statute provides a personal cause of action for you to sue somebody who impersonates you, and it includes criminal sanctions with up to a year in prison for the impersonator. Of course, you would have to convince your local

California prosecutor to enforce this statute against your impersonator before criminal sanctions could be imposed.

Privacy Laws in the United States and Elsewhere

Many countries have totalitarian histories that demonstrate the harm that can occur to individuals when personal privacy is not respected. An Italian criminal prosecution against Google and Google executives for allowing video of a sensitive matter to be posted online was based on violations of privacy rights as those rights are understood in Italy. It is highly unlikely that a company would be liable for criminal sanctions in the United States if that company took down the offensive content in a reasonable amount of time.

Although we write more about the European Union's "Right to Be Forgotten" in chapter 8, we mention it briefly here because it sets an interesting precedent, not just for its member countries but for citizens around the world. It's too early to know what the long-term impact of the European Union's decision to enforce a Right to Be Forgotten with technology companies will have. However, it's a safe bet that this is not a passing fad.

There are concerns that giving you or organizations more control of their Internet identity, under a right to be forgotten clause could lead to censure of the Internet. Free-speech advocates worry that the vagueness of the EU law could lead to pressure for all tech companies, including Google, to scrub results across the globe, delinking news stories and other information that may make you look bad, but that they consider to be news or historical data.

Here's a quick history lesson of how the European Union reached this conclusion. A Spanish citizen filed a complaint with Spain's Data Protection Agency and indicated that Google Spain and Google Inc. had violated his privacy rights by posting an auction notice that stated that his home had been repossessed. The repossession of the home had been resolved years earlier, but because the Internet never forgets, the personal data about his financial matters haunted him online.

This man wanted the newspaper, Google Spain, and Google, Inc., to be required to remove the old news so it would not show up in search engine results. The Spanish court system reviewed the case and referred it to the EU's Court of Justice. Here is an excerpt of what the May 2014 ruling of the EU Court said: "On the 'Right to be Forgotten': Individuals have the right—under certain conditions—to ask search engines to remove links with personal information about them. This applies where the information is inaccurate, inadequate, irrelevant or excessive for the purposes of the data processing. . . . A case-by-case assessment is needed considering the type of information in question, its sensitivity for the individual's private life and the interest of the public in having access to that information. The role the person requesting the deletion plays in public life might also be relevant."[17]

In the United States, federal law provides a level of immunity for companies that host online information posted by others. In 1996, fearing that online content-hosting companies such as America Online, Yahoo!, and Prodigy would self-censor Internet discussions to protect themselves from liability, the U.S. Congress passed a law to exempt those companies from liability for the information posted on their services. This exemption applied even if the information was defamatory or violated copyright or other laws, and even if the hosting company acted to manage the content on its service.

This law was passed as Section 230 of the Communications Decency Act. Although other provisions of that act were struck down as unconstitutional, Section 230 has stood against all challenges. Practically, Section 230 makes it nearly impossible to successfully sue an Internet host for publishing offensive information, protecting the wealthiest prospective defendant and leaving you with only an individual, who is sometimes anonymous, to sue for the posting.

The bottom line is that information that is covered under U.S. privacy laws receives limited protection. Once the data are released into the wilds of the Internet, you have no right and probably no ability to chase it down and have it removed. To make things worse, once information has been exposed, anyone who sees it can copy it and use it, probably without leaving a trail that you can follow.

The Role of Civil Law

Some online defamation may violate your legal rights. To address such destructive online material, it helps to know what laws apply to the Internet. Understanding reputational rights under the laws of the United States and the laws themselves, which we cover in this section, can help you to protect your reputation online.

Reputational Rights

Technology changes society faster than the law can react, so U.S. law relating to the Internet often lags behind the changes that the Internet brings to our lives. Laws protecting people's online images have therefore been slow to develop.

Some laws, such as those that protect you from defamation, have existed since the time of the Roman empire but have not yet been adapted to address the reach and speed of information flowing across the Internet. Other laws, such as those that protect the privacy of certain personal information, have only been passed recently and judicial interpretations that could help us understand how privacy laws will be enforced aren't yet available.

Laws that could be used to clean up your online persona, such as those that protect public image, do not exist in many states, and when they do exist, they

TO SUE OR NOT TO SUE?

Filing a claim in the courts may be the last resort for righting a wrong, and such an action is not appropriate in many situations because filing a suit and seeing it through is an expensive proposition all around. Lawsuits drain both the plaintiff and the defendant of money, time, energy, and often emotional well-being, because each side attacks the other and personal character is called into question in a public forum. Win or lose, you still suffer these costs, and it is rare that your opponent will be ordered to pay your legal fees. You should never rush into a lawsuit without knowing exactly what you want to get out of it, understanding that the process is unpredictable. There are at least two sides to every story, and many people cannot see the other side very well when they are angry or upset. Even if it seems that you couldn't possibly lose, you may still lose. However, lawsuits are the method that our society uses to resolve its most difficult disputes.

often apply only to the famous. In short, although several categories of law exist that might help you to manage and clean up certain aspects of your Internet reputation, taking legal action tends to be an inefficient and often ineffective solution to many forms of Internet exposure. Still, there are three areas of law that may pertain to your situation: defamation, rights of publicity, and privacy laws. Privacy laws have been discussed in this chapter in the context of identity theft, so let's take a look at defamation and rights of publicity.

Fighting Falsehoods: Defamation Law

The oldest and best established laws protecting your image address malicious publication of false information, or defamation. Defamation is the act of smearing a person's reputation with false and unprivileged statements with the intent to hurt the person or with negligent disregard for the harm being caused. *Libel* is the term for written defamation, and *slander* is the term for verbal defamation.

Winning a defamation case is a public confirmation that the statement against you was false, and you may be awarded damages or attorney's fees if you can show that the false statement was particularly malicious.

Defamation law varies around the world. In the United States, celebrities must prove that a defamatory statement was made with actual malice, but private citizens have an easier path to success. As a person who is not considered to be a public figure, it's likely that you only need to establish that the statement made about you was false, that the statement caused you harm, and that the statement was made without adequate research to determine the truth.

The defamatory statement must also not be privileged. For example, a statement made in a court filing is likely to be privileged even if it meets all the other criteria and therefore would not be considered to be defamatory. Also, keep in mind that some statements are merely opinion and would not be subject to defamation suits. If someone simply called you a "jerk" online, that's a broadly unspecific expression of opinion that probably couldn't serve as the basis of a defamation lawsuit. If the same person calls you a "criminal" or a "philanderer," then you may be able to prove that the statement is false.

Nearly all states in the United States consider certain types of malicious statements to be such clear examples of defamation that you would not even have to prove that the proclamation was harmful. These automatic defamation statements generally include allegations of criminal conduct or statements concerning a "loathsome disease," such as a sexually transmitted disease or leprosy.

The Risks of a Defamation Lawsuit

Baltimore reputation management expert Henry Fawell points out one of the rarely considered hazards of protecting your image from online defamation. By filing a lawsuit, you are publicizing the allegations.

According to Fawell,

> If someone has published defamatory information then your legal footing may be stronger, but that doesn't necessarily mean filing suit is wise. Filing suit may draw more public attention to a statement that otherwise may have sunk into obscurity over time on search engines.

Consider the case of Washington Redskins owner Dan Snyder. Snyder filed suit against the *Washington City Paper* in 2011 for what he deemed a libelous profile published by the paper. The profile was indeed mean spirited, but libel can be difficult to prove. From a public relations standpoint, the suit backfired. Before the suit, hardly anyone had heard about the profile. The *City Paper* is a tiny publication with a minuscule readership. When this tree fell, nobody heard it. By filing suit, however, the story dominated news in Washington for days. Snyder ensured that tens of thousands of Washingtonians—maybe hundreds of thousands—would read a profile they otherwise never knew existed. Once ESPN featured the story on the front page of its website, it was a national story with millions of readers.

The lesson? Before filing suit, consider all the public relations consequences.

Who Is a Celebrity Online?

It will be interesting to see over the next decade if the legal standards for proving defamation change because nearly everyone has an online persona, viewable around the world, and may be considered, in some sense, to be a public figure. The televised *Real Housewives* and the kids from *Jersey Shore* are public figures. Mommy bloggers, Wikipedia subjects, and others with a large Internet following may be public figures.

What about those of us who have Facebook and Twitter accounts but who don't cultivate celebrity? Will our newfound public images subject us to the same standards that now apply only to public figures? If a public figure is a person whose life is available for the world to see, then many teenagers with active Facebook pages qualify.

Challenges of Proving Defamation

One of the primary problems in fighting defamation on the Internet is the challenge of discovering who wrote the defamatory statement or Photoshopped a picture in a defamatory way. The law provides for civil suits against a "John Doe" whose name you don't currently know but you intend to identify later. However, because one premise of U.S. litigation is that people have the right to defend themselves, judges will not let such a case proceed very far without a named defendant.

In addition, a defamation plaintiff needs certain companies to provide information leading to identification of the "John Doe" being sued, and in most cases those companies will not simply hand over private information about their customers. You'll almost always need a discovery order from an official court case or a court subpoena to track down an information poster's Internet service provider address. If the comments were posted anonymously from a public computer in a library, then you may never track down the person who made the defamatory comment. Legal process is simply inadequate in these cases.

A further complicating factor in online defamation cases is the international reach of the Internet. Although it is likely that most defamation cases will be filed against people who know each other and live in the same state or country, the Internet allows a person in far-off locations such as Australia, Ukraine, or India to easily post a defamatory statement about you. U.S. courts are not usually interested in adjudicating civil matters against foreign nationals, and even if you were to do so in a U.S. court, a civil judgment against a foreign national may not be enforceable in his or her home country.

Unfortunately, if you chose to seek your remedy overseas rather than in U.S. courts, maintaining a defamation case in any other jurisdiction would be prohibitively expensive and perhaps even impossible because the principals have to travel too far. Also, some countries have legal systems that are notoriously protective of their own citizens or simply so inefficient that cases can take decades to reach judgment.

There's a Cost

Long before the age of the Internet, defamation cases were expensive and difficult to bring to trial. The famous sharpshooter Annie Oakley filed a series of cases against blatantly false stories in newspapers and won all of her cases but lost money in the process. Now that publications can be posted by anyone and be read by anyone with a computer or cell phone, the nature of the defamation case should probably be streamlined to accommodate the many people affected by online libel.

The barriers to publication of scurrilous lies have fallen considerably, but the barriers to addressing and refuting those claims under law are exactly the same as those Annie Oakley was forced to contend with at the turn of the *last* century. We can only hope that legislators will eventually correct this imbalance.

Contested defamation cases can cost tens of thousands of dollars in attorney's fees, so make sure the benefits of filing a lawsuit outweigh your costs. Defamation cases tend to be undertaken for resolution of reputation, and big damage awards are rare. And like many areas of litigation, wealthy defendants can drag your case out for a long time and drive expenses up. Conversely, a poorer defendant may not fight as hard but may also not be able to pay damages awarded against him.

Your Right to Control Your Own Image

Another area of the law that could be used to help clean up your online image is the law relating to rights of publicity. The right of publicity, where it is recognized, protects the rights of a person to control the commercial exploitation of her name or image.

If you see a picture of yourself online, you might wonder if you have a legal right to force a company to take your picture off its website. Unfortunately, in the United States, the answer to that question is defined by the context of the situation—primarily by whether you are a celebrity with an economic interest in your image and whether the picture was offered as part of a commercial deal to make money for someone else or to show that you endorse a product or service.

Your rights with regard to a picture posted online will depend on the laws of the state you live in and the laws of the state or country where the picture was published. In most cases you are unlikely to have a legal right to force someone to stop using your image online unless you are a celebrity, the picture is being used for financial gain, or you live in certain states that have aggressively protected this right.

There is no broad national law in this area. Some U.S. states, such as California, New York, and Indiana, have passed statutes protecting the right to profit from use of a person's image or other aspect of personality as property. Other states, such as Georgia, have extensive case law that reaches a similar conclusion. Many states have neither.

ONLINE QUIZ: DO YOU HAVE THE FACTS TO SUPPORT A U.S. LAWSUIT TO PROTECT YOUR ONLINE IMAGE?

1. If someone has written unflatteringly about you online, were those comments
   a. false?
   b. intentionally malicious or made with a reckless disregard for the harm they might cause?
   c. harmful?
   d. not stated in a formally privileged way, such as filed legal pleadings?
   e. all of the above?

2. If someone has copied your writings online, has that person
   a. failed to credit you as the originator of the comments?
   b. quoted your work as part of a scholarly publication?
   c. used the entire work or a large portion without your permission?
   d. included the writing in a news story or commentary?
   e. done all of the above?

3. If someone uses your picture on his website, are any of the following true?
   a. You are well known, and the site falsely claims that you endorse its product.
   b. You are well known, and the site owner has paid you to endorse its product.
   c. The picture shows an embarrassing use of alcohol or illegal drugs by you.
   d. You didn't know you were being photographed.
   e. Your picture is on the Facebook page of someone you dislike.

4. If information relating to your health is published online, was the publisher
   a. the local police department commenting on your condition upon arrest?
   b. a pharmaceutical company listing you as a user of its drug to fight depression?
   c. an ancestry company reviewing your DNA sample to alert you to a hereditary disease in your family?
   d. a coworker cruelly joking about your recent weight gain?

(continued)

148

e. a medical researcher reporting on testing results with your permission?

5. If your financial information is listed online, was the publisher
   a. *Forbes* magazine, listing the world's richest people?
   b. a picture of a receipt from an online merchant showing the last four digits of your credit card number?
   c. your bank, which failed to block access to lists of account numbers and balances?
   d. a contact on LinkedIn who estimates your net worth based on your employer's published statistics?
   e. the Securities and Exchange Commission describing your stock holdings in public companies?

   Answers: A lawsuit in the United States is likely to be best supported by (1) e, (2) c, (3) a, (4) b, and (5) c.

Where rights exist, they tend toward recognizing commercial protections of a persona. However, with few exceptions, these rights would not come close to protecting the right of an individual to stop his or her picture from being used if taken in a public place and for a noncommercial purpose. If you live in the United States, even if you do not like the way your image is being used online, under current law you probably can't sue the publisher of a true story, description, or accurate picture of you by claiming that the publisher violated your privacy. Rights of publicity for regular, non-celebrities simply do not stretch very far. We can only hope that, in the new online world where everybody has a public persona, the law will eventually protect us better than the current mishmash of inconsistent state rules.

CHAPTER 7
BRANDING YOUR PUBLIC PERSONA

Just as companies like McDonald's and Apple have a public image, you have a public persona for all to see. And just like these companies, it's to your benefit to be aware of your public image and to manage it. This concept is called *branding*, and in the Internet age, we all have one or more brands that represent us to the world. In this chapter we discuss how you can take control of your brand and make it work for you.

The Need for Branding

Developing your personal brand online has become so important that Syracuse University bought subscriptions to the Brand-Yourself.com platform in 2010 for all graduating seniors.[1] The company was started by a group of former Syracuse students who noticed that some students do not get selected for internships or paid jobs because of their online persona. This service hopes to put the power back in your hands to manage what you look like to others online.

Graduating students were able to use the services, compliments of the university, for six months. They could review their online persona by checking all of their social and professional networking profiles and making changes to them as needed.

In a Mashable.com article written as part of their "Real Results Series," Mashable and a company called Gist analyzed how job seekers were finding jobs by building positive online personas and using social media. One person featured in this series was Kasey Fleisher Hickey.[2] Hickey was active online, even maintaining her own food and music blogs. A recruiter saw Hickey's blog and was positively impressed by her posts and knowledge, so much so that Hickey was recommended for a job.

These are just two examples of the importance of online branding; still, when we discussed the concept of branding yourself online with people as we worked on this book, we received a variety of reactions that typically fell into one of two categories.

- Open Bookers: There is a group of people who claim that their lives are an open book, they have nothing to hide, and they will not waste their time worrying about their online image.
- Deer in the Headlights: There is another set of people who are concerned about their Internet personas but are somewhat immobilized. They feel as if their online image is out of their control, and they are therefore defeatist about changing it. If you feel this way, we have to acknowledge that a lot of the information posted about you is out of your control. You cannot control the tax records posted online. If you spoke at your kid's PTO school meeting, they might post minutes with your protests about a new school schedule.

PERSONAS AROUND THE WORLD

We're not the only ones who think it's important that you understand and control your online brand. Consider this: The *New York Times* reported that the European Union has created a campaign called "Think B4 U post!" cautioning people to think before they post about themselves and others. Also, France's data protection commissioner, Alex Turk, has asked for legal protections that allow individuals to control their online persona, asking for a legal right to "oblivion."

HOW THE WORLD CAN FIND YOU

Type "LOL Facebook Moments" into a search engine, and you'll see posts of those secrets that you, your friends, and strangers probably thought were private. You can also go to ReasonsToHate.com and see way too much personal information about marriages and relationships falling apart, new loves, and more. There is even an index to track any posts that mention "I hate . . . ," "I love . . . ," "My boss is . . . ," and other topics. Because of the ability to unearth content about you online, you need to be careful, not just about making strong privacy settings but about what you post.

No matter which category you fall into, if you need a reason to compel you to take control of your online image, consider this tragic story of a mom who discovered that someone had set up a fake profile for her son on Facebook. The kids who set up the profile hid behind her son's good name to make racist and sexual comments. The fake profile was removed, but only after the Facebook account had more than five hundred friends. Many of the classmates who were friends to this bogus profile had no idea it was fake and assumed that the nasty posts were made by the boy whose name was in the profile. The boy and his mom are concerned that during college background and college sports team checks, these ugly posts, under his name, could have an impact on his ability to join a team or get into a college.[3] So, if you hesitate to take control of your online brand for whatever reason, consider this: if this mother and son had not been vigilant and had not acted to take the bad content down, a bad situation would only have gotten worse.

How to Build Your Online Brand

So, how do you begin to create an online brand that will work for you? Your first step is to understand how impressions of you work and then create a strategy and plan to determine what brand you want to project.

Be Aware of Impressions

If you did your homework in chapter 4, you have some idea of what information is out there about you. Take a look at that information now and consider your online activities. These can provide a positive or negative impact on how people see you, and could include the following:

- clubs that you belong to
- recent events you attended
- political affiliations or events
- your relationship status
- your e-mail address and social media handles
- your online connections to schools, universities, companies, organizations, and people

Try to analyze not only what content is online about you but also what impression your activities and comments are making on others.

Have a Brand Strategy

Having a strategy is a critical part of building your brand. You can make this strategy as simple or as complex as you like based on your needs and preferences.

HAPPILY EVER AFTER?

Relationships are particularly tricky. Note that marriage and divorce records are public, and many are available online. In addition, because many property tax records are online, people can view those and make a guess at your relationship situation. Although you can't remove those records, you can be consistent about keeping your relationships private by paying attention to what you mention in your online profiles and postings.

Start by jotting down a few key words that relate to the picture you want people to have in mind when they look you up online.

Create a Mission Statement

Dr. Stephen R. Covey, author of *The 7 Habits of Highly Effective People*, suggests that you create a personal mission statement to define your life's focus based on your purpose and moral compass. Your personal brand strategy should be a complete picture of your full persona that includes the parts of you that you want to show to people. A brand strategy could include your life goals matched to what you want others to see about you online.

For example, when our fictional Bob writes his personal mission statement, he wants to come across as an experienced lawyer but also to make sure that his online persona would be someone that prospective women would like to date and maybe even marry. Bob should monitor the tweets and posts he makes on social networking sites to portray that image. If he likes to cook or golf or bowl, he could join online affinity groups and post frequently. He might meet the woman of his dreams while they are both posting about the merits of bowling at the Sunshine Lanes.

Set Goals

Next, you need to set your goals and think about how your goals and the actions you take to achieve those goals will create your brand.

If you were Oprah Winfrey a few years ago, your online brand strategy might look something like this:

- Goal: To host a dynamic and popular television show that drives viewers to a website that encourages them to watch the show.

- Goal: Produce a fast-paced and modern magazine.
- Goal: Make all who connect to the brand walk away feeling positive.

Have an Action Plan

After you have a branding strategy, before you start deleting or posting new items, take some time to think through and write up short bullet points or sentences for the various parts of your brand.

For example, following our previous example of imaginary Oprah goals, the online branding extension of those goals might look like this:

- TV Show Brand: Approachable and friendly, sitting and chatting among friends while millions watch.
- Magazine Brand: Approachable and friendly, giving advice to friends, helpful and timely.
- Overall Brand: No negative "gotcha" reporting, little focus on the sordid side of stories, "pay it forward" shows focused on compassion.

Here's a sample to guide you on how to get started.

Professional Plan

Whether you're delighted in your current job or, like Sally, you are still looking for your dream job, your career is an essential part of your Internet brand. It's important to have a résumé, but it's merely a starting point. To show what Sally is capable of, she could highlight recent projects, where appropriate, on her own blog or submit a case study paper to popular industry blogs. In the digital age of texting and tweeting, you need to also have other entry points where people can see what you have to offer. Be sure to jot down and post professional information, such as your top three skills, career-associated special interest groups, recent awards, attendance at conferences, and future goals. Note that chapter 9 discusses your professional brand in more detail.

Personal Plan

Your personal and professional life do blend on the Internet, whether you want them to or not, but your personal brand can be as important for you to work on as your professional brand. Some items that you may consider including in your personal profile are listed here:

- Hobbies and interests outside of work
- Favorite books, newspapers, or magazines
- Recreational activities

- Favorite sports teams
- Foodie, vacation spots, or art critiques

Here's a rule of thumb to use for personal posts: every post or mention of an activity outside of work should be suitable for sharing with friends, family, coworkers, and strangers.

How to Promote Your Online Brand

Think about brands and household names that have the most positive image. What do the names Disney, Ritz-Carlton, or Johnson & Johnson suggest to you? When you say these names, you probably get a specific feeling. Disney offers wholesome entertainment, Ritz-Carlton may suggest luxury, whereas Johnson & Johnson constantly promotes itself as a safety-conscious, family-oriented company. These companies have promoted a brand, and that same process can work for you.

If you come across at work as a capable professional who is focused, means business, and is a conscientious employee, congratulations! That is a great brand to own. But if, in your online life, you post negative comments about your job, you leave your boss, colleagues, and future employers to wonder which brand image is authentic.

In this section we have put together a step-by-step process that will help you manage your brand online, plan the content you should post, and decide where your brand should be located. This process is especially helpful if you are considering changing jobs, starting a business, have had a change in your personal life, or are a young adult getting ready for college or looking for work.

Own Your Name

A great way to manage what the search engines display about you is to own your own name on the Internet. Start by purchasing the domain name that matches your full name. There are several affordable options for doing this. As of this writing, some popular services you can use to purchase a Web URL include GoDaddy.com, 1and1.com, and Google.com.

You can also establish your name to set up blogging sites. Some popular blogging site options include Blogspot.com, TypePad.com, WordPress.com, Xanga.com, and LiveJournal.com. Even if you don't plan to blog on a regular basis, you can post information there about you now and then to establish your brand.

One way to create good online karma for your name is to have a good profile on LinkedIn.com. LinkedIn.com typically hits the top of search engine results. Other professional networks that you may want to explore are XING.com,

NetworkingForProfessionals.com, and Ziggs.com. There may also be professional networks specific to your industry. Figure out which ones are the most reputable and set up a profile on them. You can also track your online reputation "score" by looking at services like Klout.com. Once you have a Klout profile, this service looks at information you have posted and your connections across social networks to rate your overall influence. Expect to see more free sites that help you manage your overall online persona.

Getting involved in social networks is another great way to claim your name and promote your personal and professional brand. These networks are typically free, and the more popular ones rank high in search engine results. This means that, if someone searches for your name and you have a LinkedIn account and other social networking accounts, your posts about yourself are most likely to rise to the top of search results.

Finally, consider owning your visual brand by creating and using a consistent avatar. Avatars, which are essentially an animated version of you, help people identify a person or a brand consistently from blog posts to comments on news sites to Facebook posts. There is a free service at http://en.Gravatar.com that allows you to create an image and a profile that you can use on every site where you post information.

Start Your Branding

A 2013 *New Yorker* article, "You Are What You Tweet," still rings true. Regardless of the social platforms you use, what you post, repost, like, and link to form your brand. The top three social media sites today, based on global traffic and usage, are Facebook, Twitter, and LinkedIn.[4]

You need to set up a Facebook account right away if you want a stronger online presence because this is currently the leading social networking site. When setting up your Facebook account, you should consider your brand, and you should use settings that protect you from identity thieves, cybercreeps, and cybercriminals.

Connect with others on Facebook. For example, if you are a website designer, then you may want to follow groups that talk about the latest in Web design, graphics, and styles. You can search for and follow or friend others in your profession. You can also post pictures of your work and pictures of you at various website design events or working on a client project. All of these actions project the image of a website designer who is creative, innovative, and connected to others in the profession.

You could also become well thought of by posting helpful hints, ideas about tools you like to use, and even compliments on the work of others. As you build a following, you may find that people with whom you network online can lead you to jobs that you would not have learned about otherwise. In Mashable's article on how people are using their online persona and social networking to find jobs, David Cohen's online persona and social networking helped David to find his dream job. He tried traditional methods with little success; then he saw a friend of a friend on Facebook who had just started a new Internet marketing agency. David didn't know this person, but he took a chance and sent the agency director an instant message on Facebook. By having a positive online persona and using the social networking feature to reach out to someone, he was able to connect, get an interview with the company, and land a job.[5]

You may also create a Twitter account. Twitter allows you to create a profile and link people to any site of your choosing to learn more about you. Twitter positions you as an expert, whether related to hobbies or your professional skills. Tweeting can be a great way to connect to other people with your same interests or who work in your profession to share information and opinions.

You can also link your Facebook, Twitter, and LinkedIn accounts so a post on one feeds to the others and you don't have to do multiple postings.

Be proactive about your posts to establish yourself as a competent professional and someone others would want to befriend. You can send positive posts on the latest industry news or organizations that appeal to you. Consider making it a goal to post an idea each week that might be helpful to others in your profession, or use the post to ask other professionals for advice.

If you're a student, also consider using your university sites. Sites for current students and alumnae are increasing in popularity and are a great way to show your professional interests and to highlight your current professional status.

Use Promotional Sites

Become visible on sites that help to push your brand. Sites that you can use to promote your brand include those listed here:

- About.me, a site that will consolidate all of your social networking sites into one profile.
- Flavors.me allows you to create a one-stop site that can include photos, your résumé, and interesting information about you.
- Flickr.com is useful for photo sharing and staying in touch.
- Tumblr.com is great for sharing anything from posts to music.
- Reddit.com is a social bookmarking site that allows its user community to post recommendations and news.
- StumbleUpon.com is another social bookmarking site that allows people to share favorite links on the Web.
- Plaxo.com allows you to bring your contacts together in one place across multiple Internet services and devices, whether they are from your phone, e-mail account, or social networking sites.

Choose Locations Deliberately

Once you figure out what you want to present as your brand image, you then need to decide on your brand placement. Just as Disney and *Playboy* advertise in certain venues that match their image, people should find you in all the right places online. You need to be where the action is on the high-traffic sites but also in locations consistent with the brand image that you want to portray.

Beyond social networking sites, you have an even greater opportunity to promote your brand by becoming a guest blogger on other sites, leaving positive and thoughtful posts on discussion boards, contributing to articles that are posted online, and looking for professional affiliations that might be interested in cross publishing your blog posts. Research the various locations where your brand should appear. If you are interested in gaming, then a gaming portal might be a perfect place for you to post comments with a link back to your blog. Keep aware of your personal and professional brand: If you love to game as a hobby but your profession is wealth management, think through whether posting comments on a gaming site with links to your wealth management blog is a good fit. Based on your clientele, it might or might not be.

A fantastic resource for researching the latest trends in online social collaboration and social networking is the group Mashable at Mashable.com. This site might give you some great ideas for what is appropriate for your brand location.

Maintain Your Persona

Now that you have a strategy for what you want your brand to look like and have started to post information on various websites, it's time to think about the best way to maintain your persona.

You might want to commit to a time each day, week, or month when you will run through the steps for researching what information about you exists online, as covered in chapter 4.

If you need some help keeping track of the content that you post, in this section we propose three options that will help you maintain your chosen persona.

Set up Automated Alerts

An automated alert is a handy feature that you can set up on many accounts to alert you to changes. For example, you can create an alert that sends you an e-mail message every time someone posts your name on the Internet.

Some popular alert services are as follows:

- Google Alerts provides a service to track postings about a topic you care about, track online mentions of your favorite sports team, or to track occurrences of your name or your loved ones' names online. Visit www.google.com/alerts to use the simple interface for setting up and managing alerts. You can set up your alerts to arrive at a time interval that works best for you. Yahoo! offers a similar service at http://alerts .yahoo.com.

- Blog alerts are a great way to help you moderate comments to protect your blog's image. For example, you may find that there is one person out there who likes to post inappropriate comments. Blog alerts let you know that someone has left a comment, allowing you to approve it before it goes online. There are various blog alert tools, but one we like is Technorati. You can set up your blog on their site and ask to be notified if your name or blog posts are referenced anywhere online.

Check and Aggregate

Set up a time on your calendar to regularly check on each of your various websites, links, and social networks to see what might have been posted there. In addition, choose tools to help you aggregate and search for any posts about you. One tool for this is Plaxo, which we mentioned previously in this chapter; another useful tool is Memotoo. Memotoo syncs your phones, tablets, and larger computers so that changing a contact in one device is mirrored in all the others. It also allows you to sync your devices with social networks and accounts held on websites.

Be Active

You need to keep your information current. Many of us do not have the time to post something to a blog every day or to send Facebook posts and tweets that are intelligent and improve our brand. If you post every day without a plan, you might post things in haste that you will regret later.

Fortunately, there is an easy way to keep your online brand up to date. You can install many of the major social networking applications on your smartphone and use those times when waiting for a train or an appointment to post an image or comment to one of your accounts.

Use Fee-Based Services

Though the steps we listed here are simple, you have to find time in your busy day to do the maintenance and set up the alerts. If you find yourself in a time crunch, or you believe a loved one will not do this for his or her Internet persona, there are various reputation management services out there that you can use, for a fee.

Some popular fee-based services include Reputation.com, Defendmyname, and Naymz. They offer services and pricing plans that are based on how much you want them to handle for you and what actions you want them to take on your behalf. We recommend that you do research to stay abreast of the latest services available. We also recommend that you do comparison shopping before you sign a contract and, if possible, sign up for a trial period before you commit. If you're interested in such services, chapter 5 provides a little more insight into Reputation.com.

Back up Branding with Commonsense Behavior

There are some behavioral rules that are always good to follow if you want to have success with your online image, so in this last section we provide you with a few rules to live by.

Don't Let Emotions Rule

Traditionally, you could have a public life and a private life. Things are different in the Internet age. Your private persona tends to bleed into your public persona. Just assume, no matter what you think a Web service privacy statement says, that someday others will see that post, transaction, or information about you. During an emotional moment, one of extreme happiness, anger, or sadness, you may post something that you regret later. If you have an extreme emotion, sleep on the news or event before posting about it on the Internet.

Avoid Mistaken Identity

Watch out for mistaken identity on the Internet. There might be many other "Mary Smiths" online whose behavior doesn't match your desired image. Consider using a middle name or initial across your sites to help minimize the chance of mistaken identity. Posting an avatar or profile picture consistently across sites is helpful in avoiding confusion with others.

Practice Internet Persona Hygiene

Even when you're careful, you might post something you regret. Maybe you have an embarrassing typo in your post. Maybe you slammed the horrible service you received and you regret your tone and word choices. Even if you delete those embarrassing posts, sometimes made during lapses of judgment, they can come back to haunt you.

If you want to delete posts, read the help or frequently asked questions section for the site. Some sites provide a method for deleting content. However, keep in mind that trails and remnants of those posts may be in other locations on the Internet or copied over to somebody else's computer, so it may be impossible to permanently delete anything posted online.

There are also services out there that can help you to prevent a social or professional faux pas. Google mail has a setting called "Undo Send" that helps you retrieve an e-mail if you have second thoughts within a certain time period. The Google mail default is to send your e-mail within ten seconds of clicking Send, but you can change the timer setting for this.

When the Post Isn't Yours

If you know the person, the first logical step may be to request that he or she remove the content voluntarily. If that doesn't work, most websites give you the option of contacting them about erroneous posts that they may then remove from their sites.

In the event that the posts about you are clearly defamatory, you can take legal action, but this is not an easy route (see chapter 6 for more about this). In one such case, a fashion designer said her reputation was ruined, causing emotional and financial damages when Courtney Love posted negative comments about her. She sued Courtney Love and won the libel lawsuit. The posts have since been removed, but it was a long and costly process to resolve the case.[6]

The Short List of Online "Don'ts"

Here's a quick checklist of things you should never do online:

1. Blast an employer
2. Complain about boredom or lack of motivation at work
3. Complain about your spouse or loved one
4. Post pictures or information about friends and family without their permission
5. Get into emotionally charged arguments on social media
6. Use a fake name to post negative comments

EXPERT ADVICE: BRANDING IN THE DIGITAL AGE

David Almacy, former White House Internet director and a spokesman for President George W. Bush, was responsible for online communications strategy and the management of WhiteHouse.gov. Dave is a partner at Engage, a digital agency in Alexandria, Virginia, where he leads the public relations and media practice for an extensive list of political, corporate, and issue-advocacy clients. Previously, he was a senior vice president, digital

(continued)

media in Edelman's Washington, D.C., office. David graciously offered his perspective for this book.

In a crowded, rapidly evolving modern marketplace, it is becoming increasingly more important to tell our own story and, thanks to the exponential growth of the Internet and social media, we have more channels than ever to do so. However, content that comprises one's online presence must be actively maintained in order to effectively achieve desired outcomes.

Some refer to this process as managing a "digital footprint," which refers to how we are represented online and consists of two primary parts: online content on websites, blogs, and social media published by you; and content posted about you by others which you can't control, but you may be able to influence. The sum of these is organized by algorithmic relevance via search engines. In other words, if you publish nothing about yourself online, then you cede total control of your online reputation to others and Google will determine which content rises to the top based on clicks, relevance, timeliness, and engagement, among other factors.

As White House Internet Director for President George W. Bush, one of my favorite projects was "Ask the White House," a weekly online interactive Q&A forum hosted on WhiteHouse.gov. In addition to the value of connecting citizens with Bush Administration officials via the website, the content that was generated during the conversation assisted in clarifying complex policy positions and optimized organic search results for both the guest and the issue long after the chat had been completed.

The first step in taking control of your online presence is to commit to it. These days, anyone with a mobile device can snap photos, tweet, post status updates, or upload video on the fly. I suggest mapping out a content strategy to ensure that anything published online is not only timely and relevant to your audience, but also aligns with the image you hope to project and the brand you wish to build. For example, I usually publish content that falls into four basic categories: media/public relations trends, politics, personal life events/interests (travel, family activities, sports), and audience engagement/miscellaneous. For me, this mix generally breaks down to 30–30–30–10, but I encourage you to find the category ratio that best fits your interests.

Next, identify a few digital channels to host and manage your content. Whether it's a blog or an official website, owning an online outlet is essential. This is a place where you control all the content and can provide unique first person perspective. When writing for the Web, be concise. It only takes a few minutes to type up a couple of paragraphs to describe an event. Stick with the basics of what we all learned in Journalism 101 by answering who, what, when, where, why, and how.

When possible, include multimedia to add a little color. Visual content such as photos and video are always a plus. They add context to the story and increase the chances that your posts will be seen. Several studies have shown that posts with visual content such as photos and video consistently increase engagement rates such

(continued)

as clicks, likes, comments, and shares. When posting, choose relevant titles that accurately describe the event along with specific keywords for your images, photo sets, and video so that others can find them more easily when searching.

Be sure to leverage some of the more popular social media sites and networks that are available to you. When setting up accounts, consider reserving consistent usernames. Of course, Facebook and Twitter are among the largest of these social communities, but there are many more to explore such as YouTube and Vimeo for video, Flickr and Instagram for photos, Path for personal networks, LinkedIn for career development and networking, Foursquare for location-based event "check-ins," and one of the newest contenders to the social space, Snapchat. Many of these tools are free and offer quick and easy upload and embed features which allow for optimal online sharing. Don't be shy about cross-promoting across channels; everyone consumes information differently.

Publishing content is just the first step. By building an active and engaged online community, you will increase the chances that your content will be seen and shared. Remember, the Web is a social medium, so it is important to listen to your community when they engage with you. Respond to their posts in a timely manner and be sure to share their content in your channels from time to time, as well. In general, try to publish regularly and offer a variety of posts such as short status updates, links to blog entries, photos, news items, video, trivia, etc.

Indeed, the power of social media is changing how we all connect and communicate. Taking advantage of these online tools will enhance your ability to successfully manage your digital footprint while simultaneously allowing you to positively affect your search results, engage with interested audiences, and build online community.

CHAPTER 8

YOUR RIGHT TO BE FORGOTTEN AND TO COMPLAIN ONLINE

A s the Internet evolves, laws and regulations are also changing to reflect societal issues and problems created by new types of behavior taking place online. Never before has the world had access to statements, pictures, video, and criticism by millions of individuals who are otherwise not celebrities. The Internet provides us with places to document our lives, thoughts, and preferences online and then holds that material for an indefinite period of time, long after we might have outgrown our own postings. It also provides places where we can criticize our bosses, local building contractors, or polluters. This chapter describes some of the new and evolving law around how to manage and protect our rights in regard to new Internet information and opportunities.

Children Are Different

Criminal law in the United States treats children differently than adults. In most cases, a fourteen-year-old boy who commits a serious crime is likely to be sent to a juvenile facility with other teenagers, rather than to prison. The period of incarceration for children who commit crimes tends to be substantially less than adults who commit similar crimes. And most relevant to this book, juvenile records are generally sealed from the public, whereas adult conviction records are public and often searchable online.

Nearly all countries in Europe and the English-speaking world punish crimes differently if committed by children. These societies believe that mistakes made in childhood should not necessarily be available in public records when they become adults. Social science research has shown us that the adolescent brain is more prone to risk-taking and poor judgment than a fully developed adult brain. Teenagers are simply more likely to act impulsively, to jump into dangerous situations, and to ignore society's rules when doing so. Modern societies recognize these facts and shield adults from many of the impulsive or thoughtless actions of their childhood. If someone at your workplace committed a crime as a teenager, you probably won't know about it.

California's Eraser Law

Recently, the State of California extended this practice to hiding the embarrassing content that California teens post on social media. As described previously, social networking sites such as Facebook and Instagram are built on the piles of past information posted onto millions of websites. Traditionally, they have not allowed users to remove much historical data from their sites. However, things have changed: As of January 1, 2015, a social networking website must allow California teenagers who are registered users of the website to remove material that they posted on the website. This policy guarantees California teens the ability to erase the worst and most embarrassing items they posted on social media before those items are discovered by jobsite hiring managers, college admissions officers, or prospective in-laws.

This California Eraser Law is not a privacy panacea. It only affects certain websites, it only protects teenagers residing in the state, and it limits its protection to items that a teenager posts himself or herself. This law does not provide a right to remove humiliating pictures that classmates may have posted of the teen attending a costume party or caught making an unfortunate dance move at

FEATURED TOOL: SNAPCHAT

An early driver of the now pervasive selfie phenomenon, Snapchat appeared in 2011 as a tool called Peekaboo. A user sent a picture that lasted from one second to ten seconds and then was supposed to disappear forever. This platform encouraged shared photographs of all types, because the sharing was immediate but the history was brief. By October 2013, Snapchat was processing 350 million "snaps" per day.

Scandals followed when people realized that not all snaps were disappearing as promised and when hackers compromised the tool. But by 2014, Snapchat was already evolving into less of an intimate sharing site and more of a marketing platform, offering text and video options, along with filters and watermark logos on snaps. Advertising appeared, and Snapchat released a Discover option that allows publishers to upload content to the service. Snapchat is also measuring popularity of public snaps.

The short history of social media is a constant resource for businesses learning to "milk the herd." A company like Snapchat builds an audience by providing interesting tools and then finds ways to make money from the crowd it has attracted, thereby evolving into a different type of site.

a club. If the most embarrassing items on a young person's social media page are posted by others, then nothing can be done to force removal. The embarrassed person will simply need to ask others to remove data on his or her behalf.

Other U.S. legislatures have been less eager to provide an eraser right to their citizens, children or otherwise. Although the concept has been raised in the U.S. Congress and in some states, no additional protections seem likely to pass. In Canada, a court in British Columbia has addressed issues of requests to remove search links and left the door open to consider a right to be forgotten, similar to one available in the European Union.

The need for protection from online teenage mistakes may be less pressing than it once was. Where young people once conducted all of their lives over accessible social media, now tech-savvy teens are using new tools like Snapchat, Whisper, and Yik-Yak that make it harder to save and trace pictures and messages. They are also registering at Tumbler and JournalSpace under pseudonyms and Internet handles that cannot be easily traced to the author. According to the *Washington Post*, "College admissions officers increasingly struggle to find Internet dirt on their applicants. The percentage of those officers who say social media has negatively impacted someone's chances has fallen, in the past two years, from 35 to 16 percent."[1] Today many people are using anonymizing tools to separate the information they post for public consumption from information that could be embarrassing.

Being Forgotten around the World

This right to be forgotten originated beyond U.S. borders. Around the world, legislators and policymakers have been looking closely at this issue. But when one jurisdiction wants to grant its citizens the right to be forgotten, it must police the entire Internet. U.S. Internet companies are currently spending millions addressing the problems of applying one country or region's laws to the global world of the Internet.

The Right to Be Forgotten in the European Union

A Spanish man named Mario Costeja González fell into debt early in his life and his home was repossessed to pay the debt. A newspaper called *La Vanguardia* printed a thirty-six-word article in 1998 describing this repossession. González was upset to find that when he looked up his own name on the search engine Google, this article was a prominent link on the search page. He requested removal of the article by complaining to the Spanish Data Protection Agency and his request was denied because the article was a true reporting of facts. However, the agency granted his request that Google remove its search engine link to the article.

Google sought to overturn the decision, but the European Court of Justice held that a right to be forgotten was a core privacy right for citizens in the European Union. Google must remove links to past information when requested by European citizens, even where the link was to undisputedly true information included in newspaper articles. France's Commission Nationale de l'Informatique et des Libertés ruled that Google must remove such links on "all domain names" of the search engine, not only from searches on the European versions of Google, but also on the North American sites. Google fought this order, claiming that the French Commission has no right to expand the scope of the European Court of Justice's original opinion, but its appeal was rejected.

Now a European citizen who committed a crime or embarrassing act can have the links to news stories and other online references removed from Google searches for that person's name. In applying the right to be forgotten on Google, the courts first found that Google processes data as it proposes search results. Because Google uses a sophisticated algorithm that ranks websites and references in relation to each other, the court felt that Google had responsibility for the placement of some of those links and references. If a citizen of the European Union felt that a particular search result violated his privacy, then the court would order Google to evaluate and respect his or her request.

This book describes a number of methods to remove, mask, or minimize problematic information about you that exists online. If you are a citizen of one of the countries of the European Union, then you may have an additional option. Google provides a form on its website titled "Search removal request under data protection law in Europe." The form, available now in twenty-five languages, asks a user to select the country whose law applies to the request. If the user chooses one of the protective European Union nations, then Google allows submission of the URL for the results that the user wants removed for searches on his or her name. The user or a legal representative must sign the form.

Google doesn't promise to remove the information right away, or at all, but it evaluates the requests under the standards set by the European Court of Justice. If Google grants a request to remove a link under this regime, it sends a notice to the site that published the article that may be unlinked, so that representatives of that site can argue for keeping the link in the search result. Google weighs the arguments of both sides before deciding whether it will ultimately remove the link. A successful application will not lead to removal of the underlying article or website, but only to remove links to that article or website when somebody searches for the person in question.

Of course, not all requests are granted. For example, Google tends to leave active links to news items about public figures. But for a large portion of the European population, the right to be forgotten provides the clearest and most certain path to removing embarrassing information from personal name searches.

Source of International Conflict

This European ruling highlights one of the most significant cultural differences among advanced societies in today's world. The United States extols freedom of speech and freedom of the press as its highest values. In this view, society is better served by allowing all true information to be available, even if exposing some of that information is uncomfortable to certain individuals or groups. In the view of the European Union as expressed in proposed guiding principles as well as by the European Court of Justice, personal privacy is the dominant value and society should protect personal privacy even if it interferes with freedom of expression in doing so. In other words, in the United States, policy makers believe that future lenders, employers, and personal contacts should have a right to know about financial or legal transgressions in the past, whereas EU policy makers would rather protect a person's privacy despite the fact that masked information may be valuable to someone in the future.

These different perspectives also demonstrate one of the most significant Internet issues that remains to be decided: To what degree can the Internet rules of one country override the opposing rules held by another country? Because the Internet crosses all national boundaries, online behavior in one nation can easily cross over into another. Right to be forgotten cases have created a significant controversy over whether one country may impose its values on other countries. Can France truly dictate what Google does on its U.S.-based Google.com website, operated and regulated out of the United States but accessible all over the world?

A dictatorship might simply set up walls to keep the offending Google site from appearing in their territory, but an open nation like France is unwilling to do so. Will the United States file a protest against its ally for attempting to impose French law on a U.S. company? The conflict lines are drawn, but no conclusions have been reached.

Countries outside the European Union have not recognized a right to be forgotten. As the Internet grows and evolves, and as the volume of data about each of us increases, we are likely to see more variations of this concept, and more people who will be allowed to appeal to their country's laws to force companies to remove accurate information from the Web.

Not all European Union countries have agreed that the EU Charter provides a right to be forgotten. In the Parliament of the United Kingdom, a House of Lords Sub-Committee on EU affairs officially recommended that the UK government must advocate against any provision ensuring a "right to be forgotten" or a "right to erasure" in any updated EU privacy regulations.

WHY WE ALL REMEMBER THE DEBTS OF MARIO COSTEJA GONZÁLEZ

Through his lawsuit, Costeja González significantly changed the world's treatment of Internet search and established the first enforced right to be forgotten. Ironically, the push to remove thirty-six words in one article has led to more than eight hundred articles* featuring his case. Rather than being forgotten, Sr. Costeja González has become famous and his entire story is regularly repeated on all forms of media. This fame is evidence of an Internet phenomenon known as the "Streisand effect," in which a person's effort to remove information from the Internet draws attention to that information and leads to the opposite result than the party seeking to remove the information intended. This effect is named after entertainer Barbra Streisand, who attempted to squelch photographs of her home and drew greater attention to the same photographs.

---

* James Ball, "Costeja González and a Memorable Fight for the 'Right to Be Forgotten,'" *Guardian*, May 14, 2014.

Chairman of the Sub-Committee, Baroness Prashar, stated that the judgment of the European Court of Justice is unworkable because it forces search engine companies to decide whether to delete requested information from searches. It also does not take into account the fact that smaller search engines don't have the resources to constantly respond to removal requests. In addition, Prashar said, "We do not believe that individuals should be able to have links to accurate and lawfully available information about them removed, simply because they do not like what is said."

Right to Be Forgotten, American Style

Given the general preference in the United States to elevate freedom of expression above the concepts of embarrassment or humiliation, it is unlikely that the United States would adopt policies similar to the European Union's right to be forgotten. Without the force of law, adults living in the United States will need to find other ways to make Internet embarrassments disappear. Although many true stories or facts from your past may be indelibly written onto the Internet, there are several strategies that may help you to remove or reduce the importance of the most frustrating content.

For example, most social media sites already allow you to request removal of the items that you have posted on your own page. Facebook claims that it will always allow people to delete the content they create. Ellen Schrage, Facebook's vice-president of communications and public policy, is quoted in *The New Yorker* as saying, "If you put up a photo or a post, you always get to take it down."[2] Facebook also offers a way to mark certain posts or photos as offensive to you because they are pornographic, annoying, or because you oppose the subject matter of the post. If you note your displeasure with a post, Facebook, at its discretion, may bring the post down. The same type of complaint system is available for nearly every major social networking site, and the site will consider removing specific submissions if you find them to be offensive, obscene, or bullying. So you can remove your own postings, and you have a good chance of removing the social media postings about you if you have a strong enough reason for requesting their removal.

Other types of sites may not be as open to removal of data. Newspapers tend to only be responsive to complaints about the accuracy of their stories. News-oriented websites will not usually pull a regular news or feature story down from the Web just because you find the news story to be embarrassing or otherwise offensive. However, the interactive nature of the Internet leads to news sites that include not only the content developed by the news organization, but also content provided by online readers. Reader-generated content is more susceptible to being pulled down from the site upon complaint. For example, the *New York Times*, in its online terms of service document, states that it "encourages active discussions and welcomes heated debate on the Services, but personal attacks are a direct violation of these Terms of Service and are grounds for immediate and permanent suspension of access to all or part of the Service." If you feel that a response to one of your comments is abusive or overly personal, you can ask the editors to remove it from the comments section, and they are likely to comply.

This type of policy is the same for nearly all sites that encourage civilized discourse and discussion around a topic, from sports websites to online retailers that allow visitor reviews of products. Amazon.com encourages visitors to post reviews and even photographs and other visual content, but if this content violates Amazon's rules, the retailer reserves the right to edit or remove such content, including content that will "cause injury to any person or entity." If you feel that a post injures you in any way, Amazon has a complaint system to address these concerns, and will consider removing the offending content. CBS Sports' website includes similar rules and rights in its terms of use, as do many others.

The Internet contains a universe of stories, comments, pictures, and other information, some developed professionally, and some written or added off the cuff by viewers and browsers like you. If you find personal attacks or information among this content embarrassing, you can always ask the site to remove the offending content. The site may or may not do so, depending on your reasons for asking and depending on who posted the problem. This is as close as most adult citizens of the United States are likely to come to a right to be forgotten.

U.S. Legal Protections for Online Criticism

Although the United States may never provide a broad right to be forgotten, it still has laws protecting your Internet identity. For instance, many of us use Internet forums and comment sections to state facts or opinions on topics that are important to us. Sometimes these statements are made anonymously using an Internet pseudonym, and sometimes they are made using our real names. Either way, depending on the state where you live, you can access legal protections for stating your opinion online.

Defending Free Speech

The United States is so committed to free speech that, not only does the U.S. Constitution protect free speech from government interference, but most states have laws that protect speakers from certain private law suits aimed at shutting down criticism. Those laws are being applied by those who post on the Internet and by websites to prevent lawsuits from people and companies who feel that certain online statements about them crossed the line into defamation.

Known as anti-SLAPP laws (SLAPP stands for "Strategic Lawsuits Against Public Participation"), thirty states have legislative or judicial protections against lawsuits filed to quiet legitimate public criticism. Varying in degree and method of protection, the various anti-SLAPP laws are intended to protect legitimate protests by citizen groups and individuals against corporations or other wealthy, powerful entities that would file expensive lawsuits against them. State legislatures acted so that the rich and powerful could not silence dissent by bankrupting the dissenters.

A good example is a California case in which a nonprofit community hospital planned to build a for-profit hospital, and a public interest lawyer who opposed the action. The lawyer wrote a letter to the state attorney general asking for an investigation of the hospital's tax-exempt status because of alleged profit-making interests of the hospital's board chairman. That hospital chairman sued the lawyer for defamation and for unfair business practices. The trial court agreed that this lawsuit was brought to silence the apparently valid criticism of the public interest lawyer and granted an anti-SLAPP motion against the hospital chairman, dismissing his complaint. The ruling was affirmed on appeal, and the lawyer was granted an award of his attorney's fees and costs in the action.[3]

Anti-SLAPP Goes Online

Anti-SLAPP rules have gained new life through the Internet, particularly in the realm of online reviews. Many people, from doctors and professors to the owners of restaurants and home repair businesses have sued to silence negative online reviews. The primary tool for this is the defamation lawsuit, but those offended by harsh and hurtful online statements have also used other laws such as copyright and unfair trade practices to push back against their critics. Some of those critics and, in some instances, the website companies that hosted the online reviews, have invoked anti-SLAPP rules, claiming that the doctor or restaurateur was using the lawsuit to punish legitimate criticism and make the critics spend money defending themselves.

For example, in the case of *Ampex Corporation v. Cargle*,[4] a publicly traded company sued anonymous posters on Yahoo! message boards for saying negative things about the company and its management. According to the court opinion, the posters claimed to be former Ampex employees and published long rants against company management and poor decision making, culminating in statements like, "All in all, it was the most miserable, sleazy, cheap operation I have ever worked for," and "It was total incompetence. It was a bunch of old guys sitting around trying to make money with a new media that they didn't understand." Ampex filed a libel suit against the anonymous posters and worked to discover their identities. The defendants filed an anti-SLAPP motion in the California court, claiming that the lawsuit was simply a tool to punish them for exercising their free speech rights to criticize their former employer.[5]

Using the court case to find the name of the anonymous posters, Ampex promptly dropped the California lawsuit and filed another one in New York. The California court then dismissed the anti-SLAPP motion, too, but the California Court of Appeals overruled that decision, finding that the plaintiffs filed a lawsuit in California for improper reasons and making the company pay the Internet forum posters' costs and attorney fees. This case illustrates the fact that though powerful interests may care deeply what individuals say about them online, many state laws exist to protect people from harassing law suits aimed at shutting down criticism.

In 2015 an anti-SLAPP law called the Speak Free Act was introduced into the U.S. Congress, but it has not passed congressional votes as of this publication. Introduction of this bill was encouraged by the lobbying arm of Yelp, the website that hosts reviews of everything from restaurants to roofers. In an article in *Mother Jones* magazine, Yelp spokesperson Lauren Crenshaw said, "This issue is really one that hits close to the heart for Yelp."[6] Like Yelp, the various media companies that provide sites for public commentary and complaints have a vested interest in protecting open and honest feedback about everything from local businesses to sports figures and politicians. An entire economy has grown around those comments on the Internet, and the anti-SLAPP laws can be used to protect people participating in that economy.

CHAPTER 9

DRESS FOR CAREER SUCCESS

Your friends and family are accustomed to seeing you in casual clothes, but when you need to impress your boss or a new business client, you probably dress more formally. When we are young our online persona is generally more casual. Growing into adulthood most people need a greater focus on professional image. Revising your online image to impress prospective business clients or employers can help you harness the Internet as a vital tool for maximizing your career success and earning potential.

Many prospective customers and professional contacts will receive their first impression of you on the Internet. By carefully crafting Internet information, you can build a solid professional persona online and impress people with your acumen and expertise long before they ever shake your hand in person.

A Professional Strategy Builds Value Online

The first step in creating a productive professional persona is deciding that you want to either build a business image around your own personal image or to build a professional image that is distinct from your casual, personal online image. Some professionals wrap their personality into their profession and allow potential clients to learn important facts about them. Others display a professional image online and hide their personal information behind "friend" walls and privacy screens. Either strategy can pay off handsomely if you execute it well and the image that you create supports what you have to sell.

We asked Adrian Dayton, a social media business consultant from Amherst, New York, about the best strategies for building a professional persona online, and he said, "For the past 100 years we have had a clear line separating out our personal from professional life. This line is really starting to blur and frankly I think it's a great thing. People do business with people they know, like and trust. Letting people really see more of your personality and what you're like outside of work can really increase the chances of you getting hired. Don't overdo it however, but also don't be afraid to share more about your passions, interests and hobbies."

Dayton makes a living helping legal and business consultants build their brands, and in those businesses, incorporating personal data into a professional online persona makes a great deal of sense.

However, for many roles, like engineer, plumber, or research scientist, it may not be as important to know about a person's off-the-job personality. Prospective customers and employers would be impressed with a deep résumé, but not care much about the personal aspects of an online persona.

If you are concerned about combining the two, you can easily meld together your personal Internet image and your professional persona, but to do so, you should emphasize online those personal facts that support your professional expertise. Your Internet image should minimize the aspects of your personal life that are unrelated to business or that might reflect poorly on your character or abilities in your professional role. Postings of words or pictures that show you in moments of personal weakness may be fun to share with your friends, but those pictures could lose you business opportunities. Raving passionately about all of life's minor inconveniences on Twitter or exploring the intricacies of your failures in love and relationships on your Tumblr account may feel therapeutic but are likely to turn off people who look to you for professional advice.

Similarly, although you have a constitutional right in the United States to speak your mind on nearly every topic, taking aggressive positions on political or social issues is likely to turn away prospective customers. Even being an overly enthusiastic sports fan can hinder your professional relationships. Hardly anyone will mind if you are a vocal supporter of your hometown team or your alma mater. However, potential clients from Boston are likely to stay away if you are constantly belittling the Red Sox online, and abusive language, even in fun, can drive off potential customers.

A Winning Professional Persona

To understand what a successful online professional image can look like, take the example of management consultant David Allen of Ojai, California. He has built a brilliant online professional image that makes use of a personal brand combined with a brand plucked from his management philosophy.

How has he done this? Start with his name. The name "David Allen" is common, and you would think it might be difficult to create a brand around it, but this consultant has done it, and done it well. He has developed his own company called David Allen Company and has a website at www.Davidco.com. His site has been optimized for search engines so he can be easily found.

Although there are thousands of David Allens, when we first typed his name into Google, the Getting Things Done website was the first item listed, and nearly the entire first Google search screen was filled with links related to this consultant, including purchase pages for his books and images of him. Google also included a box describing the consultant with several pictures, a short biography, and pictures of the books Mr. Allen had authored. Other David Allens, such as famous actors, movie producers, musicians, comedians, painters, and sportsmen, were relegated to later pages.

Allen's consultancy has a professional website that serves as the center for all the other spokes in his wheel. The front page of the Allen website is concise, regularly updated, and easy to understand, and it links to a number of pages demonstrating the depth of Allen's philosophy and marketing strategy. These links not help to explore his own site, but link to other websites that sell Allen's books or attest to his skill and experience.

As well-crafted as his website is, it is not the only way he leverages the Internet. Much of Allen's most impressive uses of the Internet to create and support his personal brand and his "Getting Things Done" brand occur as supplemental uses of media. Allen writes a special GTD blog (at the branded site Gtd-times.com) to frequently update advice to people who want to hear more about "Getting Things Done," and he also publishes a blog on *Huffington Post* that links back to his own site. There is a Getting Things Done official Facebook page that links back to www.Davidco.com and to www.Gtd-times .com. There is also a Getting Things Done site on LinkedIn.com for networking with the business community. Allen's company provides a Twitter feed with short messages coming out often to update and highlight the GTD philosophy. There is even a Wikipedia page about Allen that also links to his websites. All of

FEATURED SITE: TWITTER—TWEETING YOUR LIFE FOR YOUR BUSINESS

You have probably heard of Twitter, and you may wonder how a website where people broadcast short messages about the content of their breakfasts and the difficulty of highway traffic can be useful for business. Keep in mind that Twitter is a tool, and a tool is only as effective as the person using it.

Twitter, conceived as a "microblog," has been recognized as one of the top ten most-visited websites by a well-regarded rating service.* Twitter is a social media service that provides each user with an account to publish an unlimited number of messages, none more than 140 characters in length. Given that the previous sentence was 153 characters in length, you can

(continued)

see how limiting the site rules are. Why 140 characters? Twitter founders wanted to take advantage of the Short Message Service (SMS) text feature available on many mobile phones at the time, and that service limits mobile messaging to 140 characters. Twitter also allows links to websites and photos to be sent over the service.

However, there can be magic in brevity. The tight word restriction on Twitter forces many Twitter users to pack their "tweets" with information, humor, and attitude.

On Twitter you can develop a group of followers who will receive the 140 character messages you send out. If you are writing about the personal details of your life and the tedium of your day, it is likely that only your friends and family will "follow" your tweets. However, if you establish yourself as an expert in a particular professional topic, then Twitter is an excellent tool to send your insights and new research to a worldwide group of people who are interested in the same topic. Twitter can be an effective tool for drawing prospective business clients toward you and for establishing your areas of professional interest. As you follow other Twitter users, you can build a network of people who are interested in the same professional topics as you are and who might refer work to you or teach you more about your chosen topic.

If you regularly post to Twitter, you can easily link those posts to your own website or use an application to post your tweets on your site. Many people building an online professional persona link their social media pages to their Twitter feed and then tie them all to the site that fully describes their commercial enterprise. Cross-linking can help improve the listing of your pages on search engines, and it provides new clients with different ways to learn about you and to see different aspects of your professional personality. One feature of Twitter is perfect for business. Twitter allows users to view tweets that include a word of interest. Twitter users have cleverly supplemented this tool by adding the hashtag (#) mark as the first part of any word that is considered the topic of the tweet. So, if you want to know what is being written about privacy, search for #privacy, and you will find the topic in thousands of tweets. If your business is marketing, then the #marketing tag will find other interested practitioners to help you build a network. There are third-party tools for use with Twitter, like Tweetdeck, that can make the site even more effective as a microblogging, network-building, client-finding professional tool.

---

* "Twitter.com Traffic Details from Alexa," Alexa.com, March 3, 2016.

these external links back to Gettingthingsdone.com increase the site's ranking on search engines.

Allen has created an intricate web of interconnected content—some offered through his website and some provided on social media sites, blogs, or newspaper and television website links. Allen is his business, and his business has been successfully developed around his online persona.

Expose Your Expertise

The Internet offers a practically infinite number of sites that allow you to stretch your intellectual wings and comment on the issues of the day. The smart Internet marketer takes his or her talents to the sites that customers and referral sources visit, and he or she stands out as a bright light, contributing positively to the conversation.

There is no reason on such sites to hold back information about what you do in your professional capacity or what resources you use. Anything a potential customer would want to know about your professional life should be quickly available to them on your business website or other sites. If you have testimonials from clients, post them online. If you have won awards for service or for high-quality products, trumpet the awards online. If you are highly ranked in your profession, your website should tell us all about it.

The blog has become one of the premiere business tools of our era. Blogging is a particularly good way to demonstrate the depth of your expertise, your interest and approach to your chosen field of practice, and the currency of your knowledge. The serial nature of a blog supports short posts that you can write quickly, as well as providing your reactions to the news of the day. If there has been a particularly large acquisition or a merger in your industry, tell your clients about its implications in your blog. If your business has invented a particularly creative way to solve a common customer problem, your blog is a terrific place to explain it and to demonstrate how other clients could benefit from the same creativity.

If you don't like to write, then maybe a podcast or a video blog (vlog) is the better choice for you. Try operating your podcast like a traditional radio show, and bring in guests from your company and client list to talk about the current issues in your field. A 2015 study from the Pew Research Center showed that the percentage of Americans who listened to a podcast in January 2015 almost doubled since 2008, up from 9 percent to 17 percent. The percentage listening in 2015 was up two points over 2014 levels. This medium is growing as people become more familiar with listening to streams of talk over the Internet, and it is becoming a strong tool for business exposure.

If you are uncomfortable writing your thoughts, you may simply want to turn on your computer's built-in camera and speak them to the world via a vlog.

You can post these video discussions on your website, on a YouTube channel, or on any number of specialized blog and vlog sites. You could use the vlog to tell your customers all the new discoveries that were displayed at the latest industry conference you attended. You could show them a day in your life, perfectly illustrating the services you could provide to them. Or you could bring on guest speakers to provide testimonials. High production values tend not to be important in vlogs, just personality and useful information. Vlogger Hank Green, who speaks primarily to teens and young adults about science, has reached a billion views on his VlogBrothers Channel, hosted with his literary brother John; in all their channel has earned 2.6 million subscribers.

Take Advantage of Specialized Sites

No matter what field you work in, there are specialized websites that bring people in your field together. Participating on these sites is simple and can yield significant rewards. Projecting a positive, professional persona online can take many forms, and participating in industry-oriented websites is one of the easiest and most productive. These sites can be highly targeted, drawing primarily people in your industry who are interested in your vertical market. You probably already know the most important and popular sites in your professional area. Offering your expertise and punditry on these sites can draw customers and contacts to you who would not otherwise find you online or be aware of your area of specialty.

Successful Networking: An Example

If you're a restaurant consultant who wants to help other people open and operate restaurants in a cost-effective manner, many sites can introduce your expertise to the industry. For example, the National Restaurant Association offers a site full of information at restaurant.org, including a blog called "Membership Means Business," where experts provide information to restaurant owners. Writing for this blog would expose your ideas to the association's members throughout the United States and provide the tacit stamp of approval from the National Restaurant Association.

In addition, a number of online magazines targeting the restaurant industry host websites that are hungry for content. Restaurantowner.com is a site where special contributors provide advice to people who own restaurants. The site touts its contributors as "uniquely qualified to assist existing restaurant owners and franchisees with their operating needs and growth strategies, and to assist independents in developing their growth plan to include a successful franchise program," and the site shows contributors' pictures and backgrounds. Sharing your knowledge on this site would provide you with a significant audience of owners who are interested in your services.

Restaurantreport.com offers e-mail newsletters, social media support, and articles about restaurant management, operations, accounting and finance, public relations, and restaurant design, again allowing a consultant to target the precise audience who may need advice.

Sites such as the Food Service Forum allow people in the restaurant business to come together and talk to each other. Regular contributors on these sites can develop a personal rapport with potential customers. Participating in forums or chat sites allows you to "lurk" for a while and appreciate the flow of the conversation, see what types of people are asking and answering questions, and to slowly work your way into discussions, building trust along the way.

These sites also provide the opportunity to learn what your prospective customers are worrying about in real time and to propose solutions. You don't have to be a published author or even an established blogger to highlight your expertise and meet new clients. Chatting on relevant sites allows you to write short articles and get a more intimate understanding of potential clients' needs.

Of course, the Web provides hundreds of sites where restaurant owners can contribute to discussions about dining and entertaining, thus attracting customers to their locations and providing you, as a restaurant consultant, with valuable data about your industry. Most newspapers include a restaurant and dining section, and those articles are searchable on the Web and generally include a space for comments. Many restaurant owners pay attention to the comments and criticism received online. A consultant can target specific restaurant groups and provide advice online that may lead to business.

In your industry there are probably similar Internet sites aimed at the same customers and referral sources that you want to cultivate. Look to your industry trade association—both national and local branches—for important sites that bring your target audience and you together.

Check in with the big powers in your industry. What are they doing with their own websites, their social media accounts, and their participation in trade association online activities? Review the primary vendors that serve your industry. In many corners of the business world, vendors provide sites that pull together all of the market players for comparisons and best practices. Seek out other companies and discover how they take advantage of the Internet for commercial gain.

Industry-Related Networking

Another goal of online participation on industry sites and topic-oriented message boards is to build a network of contacts. Out here in the real world, if you hang out at the golf, tennis, horse, or fitness club, you could stir up business from the people you spend time with. Visiting industry trade shows, industry-related conferences, and education events can bring you closer to another set of people who can help your career.

The online world works the same way. Your participation in a forum dedicated to your profession could introduce you to many people with companies that are potential clients and who have common interests. Using online industry sites, you can meet the movers and shakers of your professional world without ever leaving your office or living room. Offer to help others when you can, and you will quickly learn that others will help you, too. You can use good judgment when you share your expertise without giving away trade secrets. You can build your list of contacts, make friends, meet clients, and learn more about your industry at the same time. The Internet hosts so many varied industry-focused sites that you can build a business in your field using its tools.

Linking Your Way to Success

If you put enough effort into your online presence, you may earn a link from other Internet sites to your home page or your Facebook page. Although these groups will not link to every site, they often will link to paid sponsors, to people who volunteer to serve as officers of the organization, to professionals who contribute useful content such as articles and blog posts, or to people who participate in work that improves the industry's education and practices. You can easily place yourself in any of these roles with a contribution of time and effort.

Links from your industry's trade groups or online industry magazines can certainly drive traffic to your site, but they are also likely to increase your website's search ranking on sites such as Google and Bing. These search tools factor heavily into a site's popularity and the popularity of other sites that link to it. Accepting links from a major trade organization can only raise your site's ranking on a search list.

If you are ambitious about building an online persona into a successful professional marketing tool, you should consider taking the initiative to form your own industry-oriented website. The website may be the front door for a full-blown trade organization, or it could simply be a portal site that provides information helpful to everyone in the industry. If you control one of the most important sites for everyone in your industry, you develop power and influence. People would perceive you as an industry leader and come to you for help in marketing their own products and services.

Your own company's professional website may provide similar information, but if you form a separate online place that is not associated with your brand but rather serves the entire industry, you can build a credible gathering place for the whole industry. You can still drive the activities of such a site and link it back to your own, but, if you welcome everyone in, including your competitors, you can gain a different kind of respect and position.

Using Links for Success: An Example

For example, an interior designer could create a site that provides useful links and information for the whole industry. This site might include a common database of all important furniture manufacturers, wallpaper retailers, flooring stores, and fixture specialists. The designer can expect to drive more traffic through the site than for a proprietary website that advertises a single service. Prospective customers, suppliers, vendors, and others interested in the industry will visit and participate.

Admittedly, creating a general industry site adds significant work to marketing your business, but it also provides the promise of a much greater payoff.

Build Your Portfolio

For creative professionals and consultants, the Internet is a perfect place to publish your body of work, building credibility for your ability to produce creative works or deliver projects. When you apply for a job or meet a prospective client, your ticket to success is often your résumé or portfolio of work. Employers and prospects like to see your level of experience and the type of work you have done in the past. The Internet provides a publishing platform so that prospects around the world can evaluate your work and learn about your background while they consider paying you for your expertise.

When your résumé or portfolio of work is your strongest sales tool, then the Internet is a perfect place to display it. Impress people with your experience as part of their initial contact with you. Turning to our David Allen Company example, its website includes a short but powerful biography page, and this page is the second item listed on the Google search results for Allen's name. Most of his résumé consists of affirmations from others in his field. A key endorsement states, "His thirty years of pioneering research, coaching and education of some of the world's highest-performing professionals, corporations and institutions, has earned him Forbes' recognition as one of the top five executive coaches in the United States."

If your business includes a visual element such as artwork or architecture, the Internet serves as a perfect folio for your body of work. A prospective customer can view your samples, note your professionalism, and judge whether your style strikes his fancy. The Web also provides you with the opportunity to request that other sites link to your work. If you design buildings, interiors, logos, or even websites, your work is likely to be on display on the websites of your past customers. You can link to their sites to show your past work being used by satisfied customers, helping you to attract new prospects.

Leverage Your Own Website

Every business needs a home base, especially those businesses that have no office, conference room, or storefront in the real world. For an exceptionally low price compared to renting a brick-and-mortar space, you can create and maintain an online space that can be visited from anywhere in the world at any time of the day or night. You can design it in any way you choose and can pay third parties to maintain and operate the site for you.

If you are selling your services or a digitized product such as software, books, or music, then your online storefront can become a one-stop shop for your customers. You can automate every part of your sales, marketing, and service processes.

If, on the other hand, you are selling hard goods, such as handmade purses, ball bearings, or fireplace grates, then your website can become the front office for your order fulfillment operation.

No matter what profession you are in, build a website to tell your story to the world. The site doesn't have to be expensive or complicated, but not having an online presence is like not having a business phone. Some professional sites are little more than billboards along the electronic highway, simply providing basic information about companies. This type of simplicity can work for many professionals. Professionals need a central site where their clients, prospective clients, or future employers can learn all about their relevant skills and experience. Your site can be a type of multimedia résumé, including videotaped speeches or award-acceptance ceremonies. You can include audio files highlighting your deep understanding of the crucial problems in your industry in addition to written content. Even the simplest of websites can anchor the marketing for your professional career.

Certain companies, like manufacturers that sell to a limited number of industrial clients, may also benefit from a simple billboard website. The companies only need a place that their few clients or prospective clients can find basic information, make easy contact with the company, and see the product options. However, even this straightforward and clean site can be a useful sales tool.

Of course, the Internet allows much more interactivity than a basic site might have, and your business would be wise to take advantage of these features. Many professional sites encourage feedback from customers or clients, providing customer service right from the Web page. For example, if you've ever visited certain technology sites, you know that you can request and receive support by live chat with a qualified technologist. A separate chat function window will appear, and the customer can ask questions and receive answers in real time, rather than sending an e-mail and waiting for a response or dialing into a telephone call center. The same technician may provide the help in all three instances, but a customer who can deal in real time with a representative from the company website is likely to be impressed with the level of service offered.

WHERE TO BUILD YOUR PROFESSIONAL SITE

Although you might purchase, operate, and maintain a network server in your basement, it is much more likely that you will pay a service to host and manage your professional website for you. Many companies, including Network Solutions, GoDaddy, 1and1 Hosting, HostGator, and MidPhase, offer low-cost, professional Internet hosting solutions.

When choosing an Internet host, think about price and whether the company successfully hosts thousands (or even millions) of other sites. Consider security and backup capabilities. What interactive options does the hosting company offer, and how easy will it be to add those options to your site? Such options might include a chat function to talk directly to your customers or an e-commerce feature to process sales through your website. Confirm that your hosting company provides well-designed, cost-effective options for these functions.

Ask how many clients the host works with that are roughly the same size as your business. A hosting company that primarily services Fortune 500 businesses may not be responsive to your service requests, whereas one that counts small businesses among its customers may not be able to service your company when it grows in size a few years from now. How much more will you need to pay if your professional site's traffic increases tenfold?

A good hosting company should offer website analytics that tell you as much as possible about the customers that visit your site. The simplest analytic tools from GoDaddy boast that "more than 30 reports tell you everything from how many people visit and what paths they take through your site to where in the world they live." Most good Web analytics tools offer graphics that make the numbers easier to understand.

Additional services can include dedicated servers for your site, extra security services, and search engine optimization tools. You can add software or outside services for many of these functions, but good hosting companies may offer them as part of the service.

These same chat tools are used more aggressively by some companies to spur sales. We have visited some professional sites where, as soon as the prospective customer enters the company's site with the browser, a chat window pops up, offering a greeting and help navigating the site. These chats allow a personalized experience on the professional Web page and the opportunity for targeted selling of products or services, even the ability to offer discounts that keep a hot prospect interested in making a purchase. Website analytics tools can tell your business what pages trigger the best reactions from the most prospective buyers, or at what point in the purchasing process a prospective buyer is likely to turn away. You can use this information to offer incentives or help at the precise points where a buyer needs encouragement.

You can offer your site in German, Portuguese, or Mandarin to spur international sales and signal your interest in selling within particular countries. Translation programs are inexpensive and easy to use.

Community Building

Another useful form of online interactivity is the community-building structure that allows your business's clients to connect with each other. Comments that others post on your blogs or in specialized discussion groups, as well as customer reviews of products or services, can be monitored by the experts within your company. These tools help you to understand what your customers want while allowing prospects to develop a deeper bond with your organization and a better understanding of your company's philosophy and service principles. This type of tool can be risky, as unhappy customers may try to influence others, but the best way to combat this problem is by keeping the customer chat feature in a premium section of the website where people are so committed to the brand that they pay a monthly fee to receive access to more information. Unhappy people will complain more if it doesn't cost them anything to do so. Also, a properly written Terms of Use document for your website should allow your business to edit or refuse to publish certain types of comments.

The most common and productive form of customer interactivity on the Internet is called e-commerce. With e-commerce, your professional site can serve as a virtual storefront, selling your wares without the need for a brick-and-mortar location.

Amazon.com is primarily an e-commerce company that sells just about every type of goods. An e-commerce site is not the kind of operation that an Internet business beginner should create himself or herself. Many e-commerce support companies exist on the Web to help businesses like yours with secure acceptance and protection of your customers' personal information and with the processing of payment transactions. Your business's Web developer or Web-hosting service is likely to perform these functions for a fee or will have recommendations for companies that specialize in this aspect of e-commerce.

On the Internet, your costs are much lower and your reach is much wider. E-commerce applications can be as simple as a sales tool that allows customers to buy one item at a time and may be as advanced as Amazon's one-click purchasing applications or intricate shopping cart features that allow customers

MAKING THE MOST OF A WEBSITE

It's clear that the consultant and author David Allen built a Web strategy that optimizes the way search engines find his information, so that his site is easily discovered above all the Internet noise.

The Gettingthingsdone.com page is also a model for understated branding. Its home page focuses on the message that the company wants you to hear, showing how to move from being uncertain before using the Allen system to "Ready for Anything" after getting his help.

The site addresses the two target populations that would benefit the most from his services—organizations and individual entrepreneurs.

The website lists upcoming seminars, and it also includes free podcasts and a description of the Mastering Workflow Series that users can access for a fee, bringing them exclusive products not available to the general public. These services draw people into the site and provide a deeper interaction with the Getting Things Done methodology.

All of these features create a community of Allen acolytes and clients who not only pay money to the company, but as committed and paying community members, are also more likely to return frequently to the website, and therefore are more likely to buy other products and services from Allen.

Allen also includes his other trademark brand on this site—Getting Things Done, described on his website as a "groundbreaking work-life management system by David Allen that provides concrete solutions for transforming overwhelm and uncertainty into an integrated system of stress free productivity."

Allen wrote a book called *Getting Things Done: The Art of Stress Free Productivity*, which he sells on the e-commerce portion of www.Davidco .com. His site is full of the "Getting Things Done" brand, which also carries over to seminars Allen presents to businesses. Davidco.com provides a self-testing tool to find out how good you are at getting things done and how much you might need help from Allen. His website offers the "GTD" system, including compact disc training modules, books, and a brief online membership. His company provides Getting Things Done coaching and trainer certification and free short video instructions. His online learning center is also branded as GTD Connect. His site offers links to dozens of podcasts—audio files that allow you to listen to lectures and interviews on your computer so you can hear Allen tell you in his own voice how Getting Things Done can change your business and your life.

(continued)

The David Allen Company has apparently availed itself of nearly every type of online marketing strategy. Yet this entire Internet empire seems to be based on one primary website. All the other Web tools—blogs, videos, audio sessions, social media, Wikipedia—arise from or link back to this core site.

to buy many items now or save some for later. Most e-commerce service providers offer a wide array of service and pricing to meet your company's sales needs online. Some even offer fulfillment of your product orders.

And, of course, turning your company's website into an e-commerce shopping site is the easiest way to connect to your customers and begin making money from the relationship.

Whatever functionality and details your primary website contains, it is your best opportunity for building your personal and professional brand. You can include pages of descriptions of your philosophy, services, products, links to your favorite sites, and lists of recommendations and product reviews from happy customers. Your professional site is home base for your company. Make it as simple or elaborate as you need to support your online image.

Getting the Most from Professional Networking Sites

Your professional network may be your most valuable resource. Your business contacts can connect you with vendors or customers, warn you away from faulty service providers, and help you make the contacts you need to succeed in your career or grow your business.

A well-known aphorism in the technology industry called Metcalfe's Law states that "the value of a telecommunications network is proportional to the square of the number of connected users of the system." Metcalfe's Law sums up the concept that each person you add to your network brings a value larger than himself to the network as a whole. A network of sixty-four people is much more than four times as valuable as a network of sixteen people because of the corresponding increase in the number of their contacts. The value of a network is always greater than the sum of its parts. As you build online professional networks, remember that their value grows immensely with each new person that is added.[1]

OPEN AN ONLINE STOREFRONT

If you're not prepared to operate your website as a full-fledged e-commerce store, the Internet offers opportunities to join a "virtual mall," a website such as Amazon where people gather to shop and where they can find your products. One of the oldest and best examples of this kind of online shopping mall is eBay, an online auction site that allows individuals and businesses to sell almost anything. People often visit eBay to shop for collectibles and hard-to-find items, including large-ticket items such as cars, industrial equipment, furniture, and computers.

eBay now offers both competitive auctions and fixed-price storefronts, but it also allows thousands of individuals and large companies like IBM to reach a larger and often specialized audience for goods. Whether you are selling mid-twentieth-century tiki torches, antique dolls, bronze busts of famous poets, or vintage car parts, eBay will serve as your storefront and will include your goods in user search results.

Amazon.com also offers similar services to sellers of books, music, or other goods available at Amazon stores. The primary differences between Amazon's retail hosting model and eBay is that Amazon offers fixed prices from all of its retailers, and Amazon may compete against its hosted retailers with products of its own. Amazon is one of the world's largest retailers in its own right. Conversely, eBay does not offer its own products in competition to its hosted retailers, and eBay sells many of its goods at auction, allowing buyers to bid against each other for a fixed period of time, with the highest bidder committing to purchase the item at auction.

These sites and others like them charge fees to sellers who offer goods through the various sales sites. For these fees, the sellers are entitled to list and sell items; to use the site's e-commerce software platform, including the financial tools; and to receive consulting and assistance from the site's experts in maximizing sales and shipping products. Most of these sites also build communities of sellers who talk to each other online and share tips, tricks, and traps of the online selling business, as well as sophisticated analytic tools that can help sellers explore and understand their sales figures better.

eBay's Online Commerce Machine

Started in 1995, eBay became an early Internet success story by offering a place for collectors to meet each other from any corner of the globe and

(continued)

to tap into a worldwide marketplace of one-of-a-kind items. If you were a collector of toy train sets in the early 1990s, you could find some items for your collection from manufacturers' catalogs or you could physically travel to a convention and swap with others who shared your interest. By 2000, eBay allowed you to chat with other collectors and find an entire world of rare items for your collection without ever changing out of your pajamas.

Now a hugely successful public company, eBay has purchased the PayPal payment system used in so many of its transactions, and subsequently spun off its financial businesses into separate companies. eBay expanded with a new and used car, boat, and motorcycle market, and it has added classified ads to its site. They offer a fashion section for new clothes as well as vintage velvet jackets and period dresses. The company moved its site into the mobile market, with sales applications for iPhones and smartphones built on the Android platform.

For sellers, eBay charges fees to list each item and fees when the item sells. eBay also encourages buyers after a purchase to rate the seller in several different categories. The company provides sellers with advice on growth strategies, best practices, search visibility, and selling tools, as well as an entire shipping center and a center for selling to business. In 2008, both eBay and Amazon boasted more than 1.3 million seller accounts worldwide.

The Value of Internet Networking

The Internet offers one of the best networking platforms devised by humans. Geography is irrelevant on the Internet. You can find experts in your field anywhere on earth and enlist them to help your business or career. Similarly, timing is greatly diminished as a barrier to business because the Internet allows you to work from anywhere you want, so office hours are not critical. Many Internet tools such as e-mail and messaging functions on websites allow you to connect with colleagues at their convenience.

Digitized contact lists make your professional network searchable and organized into useful categories. New mobile applications even allow the participants in your professional network to keep track of each other's locations at any given moment.

Social media guru Adrian Dayton advises:

To establish expertise online there are a number of things you can do. You can write online bios that demonstrate not just a specific practice area but that also show you to be exceptional and that highlight unique experiences that separate you from all other competitors. You can write articles, blog posts and commentary on a regular basis so that when people do a Google search for your name they see the myriad of articles that you have created. Most importantly, engage. Don't share your articles and sit back and wait for the phone to ring. Reach out to reporters, reach out to other influential bloggers and build relationships. There is not a Google algorithm that determines whether you have expertise, it is determined by people. The more influential users of social media you get to know online, the more likely it is that you will become known for your expertise. Reach out directly to ideal clients and share your content that may be of use to them.

Dozens of Internet sites are vying for the right to host your primary professional network. Some of these sites, such as the BNI network and the Rotary Club, are traditional networking organizations with real-world counterparts that are building international connections on the Web. Others, such as Facebook, Google+, Yahoo! Groups, or NetParty, are social media sites that have migrated into the professional world and want to attract customers to companies using social media.

Some social media sites were created with the sole task of developing business networks, like LinkedIn, Gather, HubSpot, and Networking for Professionals. Some business social networking sites are specialized, such as Tweeko for the technologically savvy professional person, Sphinn and Pixel Groovy for online marketers, and XING and Small Business Brief, which aim to serve the entrepreneurial crowd.

Some well-known career matching sites are more than just job-posting boards. Sites such as Monster.com and Careerbuilder.com also provide places to interact with others in your industry. Some of these sites are accessible by "invitation only," but most of them allow anyone to join and test the waters.

Choosing Your Networks

The first step in harnessing social media for your career or business is to pick one—or several—site(s) that are the best fit. What kind of a business network do you want to build, and how do you intend to use it? You may want to find a group of professionals with whom you can bounce around ideas for your company. You may want a broad group to serve as a safety net when you need to look for a new job. You may want like-minded professionals of differing expertise so that you can build a robust talent pool to pitch clients for projects.

Some sites are better than others for each purpose, so visit them and study what goes on and who is active there.

Working the Networks

Add links liberally to your professional website or to your profile page on social media. Many of these sites work by allowing people to invite others on the site into their network. The more networks you tap into, the more valuable each of your contacts becomes. Many business people see links within their market as a reciprocal proposition. If you link to their site, they will link to yours. Remember that search engine algorithms take the number and quality of links to and from your site into account, so be generous with your links.

FEATURED SITE: CONNECTING THROUGH LINKEDIN

The best-known business-focused social media site in the United States is LinkedIn, founded in 2002 and now claiming more than 332 million registered users.[*] During the initial writing of this book, LinkedIn held an initial public offering that marked its value at roughly $8, or 520 times the company's earnings.[†]

LinkedIn provides business people with several types of accounts, one free and other paid premium accounts with additional benefits. The site encourages registered users to list their résumés, along with school and work history, so that each user can be linked to classmates and current and former colleagues to build a vast network of professional contacts. LinkedIn makes it easy to invite these contacts to join your network and then recommends other people who seem like natural fits into the same networks.

When you register on the site, LinkedIn allows you to set up groups so that people in various networks can express news and opinions to those with shared interests. Group titles include such specialized functions as Worldwide Privacy Professionals, .NeT People (Microsoft Professionals), the India Leadership Network, and the Society of Emotional Intelligence Network. These groups bring people together by interest area, rather than personal knowledge, and provide a place to learn about your chosen field from other leaders in the field.

LinkedIn provides a downloadable toolbar for your browser so that you can stay connected all the time, and it offers a mobile application for use on smartphones. It also provides tools that connect directly into Microsoft Outlook, so your professional contacts can import easily into the LinkedIn social world.

---

[*] http://www.statista.com/statistics/274050/quarterly-numbers-of-linkedin-members.

[†] Erica Alini, "So, What Exactly Is LinkedIn Good For?" MacCleans.ca, May 30, 2011.

Keep in mind that you are not just accepting an individual into your network, but you are moving a step closer to his or her entire network as well. When you need information, customer contacts, or job ideas, these vast, extended networks could be helpful. If you can ask your entire LinkedIn network, and by extension each of the contacts in their networks, for the name of somebody who works in the Xbox division of Microsoft or in the legal department of the Federal Trade Commission, it's likely that someone in that extended group can help you. Clearly, when looking for a job, your extended network is useful as well. Build your network of people who work in your industry and who know others in the industry as well. This kind of network can keep you apprised of job openings before they are made public and cast a wider net of opportunity for you.

The next lesson for using your networks is to actively participate in the sites you choose. Continue to post interesting comments and findings in your area of expertise. Start a new interest group for mobile application development or medical practice marketing and see who joins and participates. Feel free to participate or moderate these discussions in topics that relate to your business or career. Make suggestions to friends looking for work or in need of assistance. The rest of your network will appreciate your willingness to help. You can use these sites to keep yourself front of mind for your professional contact group and impress them with your energy, interests, and intelligence. Smart online participation in a business-oriented network can turn network participants into friends, customers, referral sources, employees, and colleagues.

Building your online professional persona is much easier if you take advantage of the tools available on the Web. Business networking sites provide an inexpensive and easy set of tools to build and grow professional relationships and to shine as an expert in your field.

Conclusion

Your online persona is not simply an attic to be cleaned and managed occasionally. It has the potential to emerge as a valuable professional tool for you to attract new workers, impress colleagues and employers, and display your intellectual and other wares for customers. You can even use the Internet and e-commerce tools to build your own online business. Actively participating on the Internet in thoughtful and creative ways can help you to build a strong business and to put forward your best professional image.

CHAPTER 10
DON'T FORGET THE KIDS

Throughout this book we've provided information and advice about how your virtual persona can work for or against you. As an adult, you can take charge of your online reputation and even use it to be more successful in your career.

Children face some of the same challenges, but they have unique vulnerabilities. Also, today they are online in a big way, building up a reputation that will have an impact on their schooling, relationships, and careers in a manner that our world has never experienced before. Your online life may have been brief and easy to clean up; their online lives start before they are born and build up for years before they reach adulthood.

Your kids' data are considered a valuable target too. Marketers would love to know them better. Technology companies such as Microsoft, Apple, and Google provide ways to reach them at home and at school. It was recently reported that more than fifty million teachers and students use Google programs at their schools around the globe.[1]

Hackers and other parents may check out your kid online. Unfortunately, fraudsters and predators are also out there.

In this section we look at some of the differences between kids and adults, both in how each views online technologies and how they are exposed online.

Digital Natives and Digital Immigrants

If you worry about your kids online, you're not alone. According to the Pew Research Center, one in three parents are concerned about their children and their digital habits.[2]

The first thing to understand is that kids online are different from you and me. If you were born before 1993, you are considered by most experts to be a "digital immigrant." If your kids were born after 1993, they have been labeled "digital natives." The generation born after 1993 most likely cannot imagine a world where they would have a question and could not use their smartphone to look up the answer. They cannot fathom living without the Internet, where they have unfettered and instant access to information, answers, reservations, shopping, and the ability to catch up with friends almost any place, any time.

Digital natives use phones as mini supercomputers, and sometimes they even make calls on them, if they must. A recent study shows that 88 percent of U.S. teens have access to a mobile phone. Teens are opting to use a smartphone instead of a computer; in fact, 91 percent of U.S. teens go onto the Internet via a mobile device. Ninety-two percent of U.S. teens go online daily, with 24 percent admitting that they are online "almost constantly."[3]

Only 11 percent of teens use e-mail to stay in touch, and they like using a landline phone even less.[4] Their preferred method of communication is texting, followed by social networking, and then actually making phone calls on their cell phones.

According to Nielsen in a cell phone study they conducted, nobody sends more text messages than girls aged thirteen to seventeen. They send and receive an average of 4,050 texts per month, or roughly 135 to 150 per day. The biggest voice usage is by adults aged twenty-five to forty-four years of age. Kids are texting, whereas adults are e-mailing and calling.[5]

For the digital immigrant, a phone is what you use to call people, and it has some other neat features. Digital immigrants may find it tough to explain why they have concerns about what the digital natives in their households share and do when they are online. No wonder we have a hard time getting our message across—they never get the message!

Digital natives pick up technology quickly and are adopting technology earlier in life. AVG conducted a study polling 2,200 mothers with children between the ages of two to five years old across several countries.[6] The households in the study all had Internet access. For the kids in the AVG study, 69 percent of the kids were able to use a computer, and 58 percent could play a computer game. Looking at the same group of kids, only 20 percent could swim and only 52 percent could ride a bike.

Many digital natives spend most of their time, when they are not in school, online. One *New York Times* article about a study conducted by the Kaiser Family Foundation had a headline that read, "If Your Kids Are Awake, They're Probably Online."[7] The results of the study actually shocked us and confirmed a growing trend in which kids from ages eight through eighteen spend up to 7.5 hours a day connected to a phone, computer, television, or other electronic devices.

According to one source, 87 percent of young adults ranging from ages 18 through 34 who own a smartphone admit they hardly ever disconnect from their phone. Four out of five responded that the first thing they do every morning is to reach for that smartphone.[8]

A fun way to check in on your usage and compare it to your kids' usage is to try this app: http://www.checkyapp.com. The Checky app works in a similar way as Moment and Breakfree. All these apps are designed to show you how often you check your cellphone to help you create healthier digital habits.

In a presentation at a grade school, we asked a group of fourth-grade boys what they wanted to be when they grow up. The enthusiastic answer, in unison, was "a professional YouTuber" referring to individuals who spend most of their waking moments on the Internet. These folks create videos about their interests, such as baking beautiful cakes or playing Minecraft, and they are paid to do so by advertisers. Try to beat that on Career Day at schools.

So, What's the Harm?

Why should you be concerned about this extreme usage of the Internet? Every time your children go online, they create digital footprints that lead anyone, such as their friends, potential mates, college admissions, employers, cybercreeps, and cybercriminals to their digital persona. Every photo, video, comment, text, and e-mail is likely to be stored online or offline and tied to them indefinitely.

A poll of tweens and teens found that four out of ten kids regret something they have posted online. One out of three admitted that they share information online that they would not share in public. To top that off, 62 percent of kids lie to their parents about what they do online. With statistics like these, the numbers indicate that mistakes are being made by lots of kids online, including good kids with good heads on their shoulders.

If you are a digital immigrant, you are probably worried that your child's life is now public on a digital billboard advertising both their wonderful activities, but also their mistakes, for all to see. You are right to be concerned. The good news is that there are things you can do to help your kids build an online persona that is positive while also protecting them from prying eyes.

For parents with newborns on the way and infants, you cannot start too early thinking about your child's online persona. Some kids have a digital persona before they are even born. According to an international survey of 2,200 mothers, 81 percent of the kids surveyed had some online presence that ranged from photos on a photo-sharing site all the way to their own domain names and social networking accounts. Some 33 percent of U.S. mothers in the survey said they are sharing their prenatal sonograms on the Internet. Kids are making tiny digital footprints while still in the womb, thanks to enhanced ultrasound technology and photo-sharing sites such as Snapfish and Shutterfly and social networking sites such as Facebook, Instagram, Pinterest, and Google+ that encourage photo uploads. The Internet now offers today's digital version of yesterday's baby book.

FROM DAY 0: PROTECTING YOUR CHILD'S ONLINE IDENTITY

Here are three key types of information you should always keep private online.

1. *Physical Characteristics:* Do not chronicle online any personally identifying characteristics that can be used to target your child or his or her identity, such as unique physical traits such as birthmarks, the child's full legal name, time and date of birth, genetic health issues, and weight. Save those details for the paper announcements, face-to-face conversations, or the offline baby book.
2. *Genealogy:* Be careful about how much of your family tree is available online. It's public information, but you don't want to hand over your child's full genealogy to the anyone on your Facebook page. Information such as mother's maiden name or grandmother's first name is used all the time to provide access to password-protected accounts.
3. *Unique Identifiers:* Never e-mail your child's name, date of birth, and Social Security information in a single message or include it in an attachment to an e-mail. If you must share this information, use the phone, or break the data up into several separate e-mail messages sent from different accounts. Yes, the data can still be stolen, but these two methods provide a good measure of security.

As a parent, you need to take steps now, whether your baby is coming soon or already here, to inventory what you are creating to ensure it is a digital persona that your child will be proud of when he or she grows up and that keeps your child safe today and in the future.

Here are a few ways to begin to get a handle on what your child's online persona is today.

- If your child is already online, take a moment to go to your smartphone or computer, and fire up your favorite search engine such as Duck-DuckGo, Google, or Bing.
- In the search box, type your child's name and take a look at some of the results.
- If you load McGruff Safeguard on your child's computer, you can use a browser to see every website he or she visits. You can also see

what photos your child is uploading and what social media comments they are posting. The software also allows you to block inappropriate websites.

Billboards on the Internet Highway

Because kids do different things online, it's important that you understand how they expose personal information and how that information builds up over the course of their childhood and teenage years.

How Information Gets Exposed

The founder of Facebook, Mark Zuckerberg, said: "People have really gotten comfortable not only sharing more information and different kinds, but more openly and with more people. That social norm is just something that has evolved over time."[9] This is the person who was one of the great minds behind the social networking service your kids either use now or will use in the near future.

Your kids would probably not walk up to strangers and tell them all the personal details of their life and your family, but they might be inclined to post those details on social networks to their circle of friends. One click and the rant or personal thoughts and fears become digital, and digital can be forever. This generation feels that publishing personal information online means that they are being open and honest. They don't realize that they may not agree with their own posts in one week, one month, or one year from now. Still, those posts could be tied to them forever.

Your kids have many ways to expose their information, including their Internet browsing habits, text messages, online journals, blogs, posted comments, and photo- and video-sharing sites. Devices they can use to post such information include video cameras, computers, smartphones, and tablets. New services and technologies that trade in personal information are coming on the market every day, and you can bet your kids will adopt them with lightning speed.

Today's kids, because of their online habits, are heavily tracked by companies and advertisers. Your kid's favorite websites might actually be the biggest offenders of Internet tracking. The *Wall Street Journal* looked at fifty sites that kids love to use to see if those sites tracked kids and their browsing habits. When they finished their tests of the fifty sites, they had a staggering 4,123 pieces of tracking technology on their computer. The tracking tools did not collect the names of the kids, but they did collect or estimate age, general locale, preferences, and even potential ethnicity.[10]

The website with the most tracking files? Google. When the *Wall Street Journal* did their study, they found that Google was able to indicate the favorite sites of a ten-year-old girl. They showed her mom that Google pinpointed her daughter's interest in pets, photography, virtual worlds, and online give-aways such as free screensavers or wallpaper. The mother was surprised, and the little girl exclaimed, "I don't like everyone knowing what I'm doing and stuff."

Google was recently questioned about their in-school privacy processes when kids and students use their Google platform in the classroom. You may not know this but Google recently surpassed other tech companies in the classroom with Google Chromebooks representing 50 percent of tablet sales. Google also has a teaching platform called Google Apps for Education (GAFE) Core Services. The services include Gmail, Google Calendar, Google Classroom, Google Drive and Docs, and Google Hangouts. Google has signed a student privacy pledge and offers schools ways to customize the privacy features to protect student identities. The point is, the schools need to opt into these features. You can read more about Google's commitment to student privacy on their education blog at: http://googleforeducation.blogspot.com/2015/12/the-facts -about-student-data-privacy-in.html.

After teaching Internet safety classes for kids in grades K–8, we have learned firsthand that many kids have knowledge of how to protect their privacy and security online. The problem is that they only use that knowledge to block out some of the adults in their lives, such as teachers, coaches, and parents, but not strangers. Consider these statistics: more than 50 percent of kids have admitted to friending someone they did not know. Roughly 20 percent of young people using Facebook are considering "unfriending" their parents or leaving Facebook and using another social networking site.[11]

Information Builds Up

Studies have proven that parents are the first line of defense for keeping kids safe online. Still, talking about the Internet might be challenging because the concept of "forever" is lost on many of today's digital natives. Just as they can't believe they'll ever turn thirty, they don't think about what would happen if today's post turned up in their own future child's search results twenty-five years from now. As we mentioned previously, every post to the Internet creates digital footprints that your children leave behind. But unlike footprints in the sand washed away by the tide and time, just because you cannot see their posts doesn't mean they're gone. A picture or status update your child posts on Facebook may get deleted, but somewhere out in cyberland, it still exists. It may reappear when your child least expects it and have disastrous results.

As evidence of how valuable the treasure trove is considered to be, look at some of the alarming details of the VTech hack. VTech provides learning toys for kids. Hackers broke into VTech's systems and ran off with the information of roughly 6.4 million kids, including names, gender, home addresses, ages of the kids, IP addresses (which can be used to guess their geographic locations), and pictures of the kids as they used their devices. HaveIBeenPwned says this is currently the fourth-largest consumer data breach in 2015.[12]

Old-Time Rules That Still Apply

As a parent, you may feel overwhelmed and not sure where to begin. Most of us did not have the Internet to worry about when we were growing up, which makes it hard to talk to kids. Many parents tell me they have not even broached the subject because they themselves are not actively online other than, perhaps, using e-mail. The good news is that several old-school life lessons that were taught to our parents, their parents, and to us are still applicable today both offline and online.

HOW TO BEHAVE ONLINE

Six "old school" rules still provide a commonsense guide to how we should behave on the Internet today.

1. Don't post, text, or e-mail in anger. *Old School:* "Whate'ers begun in anger ends in shame."—Benjamin Franklin, *Poor Richard's Almanack*, 1734.
2. It does not take much to ruin your online reputation or somebody else's. *Old School:* "Little strokes fell great oaks."—Benjamin Franklin, *Poor Richard's Almanack*, 1750.
3. If it sounds too good to be true, it probably is. *Old School:* "Distrust and caution are the parents of security."—Benjamin Franklin, *Poor Richard's Almanack*, 1733.
4. Do not share personal information online—ever. *Old School:* The phrase "loose lips sink ships" was made famous during World War II on posters created by the United States and England to warn soldiers and family not to discuss any military matters such as departures, arrivals, or movements.
5. Everyone online is a stranger—even people who act as if they know you. *Old School:* We can all recall parents, grandparents, and teachers telling us at an early age: "Don't talk to strangers."
6. Online posts should be supportive and not destructive. *Old School:* If you can't say anything nice, don't say anything at all.

Cyberthreats and Cyberbullying

When we ask third-, fourth-, and fifth-grade children during our Internet safety class, "How many of you know someone who has been bullied online?" it almost always happens that the entire class, or most of the class, raises their hands. When we ask them to share stories without naming names of the victims or bullies, most situations turn out to have been managed well by the kids but still awful for all involved.

Cyberbullying is part of kids' lives, and both victims and bullies suffer.

Cyberbullying Defined

According to the National Crime Prevention Council, cyberbullying is "using the Internet, cell phones, or other devices to send or post text or images intended to hurt or embarrass another person." Cyberbullying can consist of anything from sending mean text messages to deliberately spreading rumors online. It may involve photos taken or posted without permission to embarrass a person. It could also involve sending repeated threats via texting or posting on the Internet or making fun of a person while encouraging others online to join in with the bullying. More than half of teens say they have been bullied online. Of those bullied, more than half do not seek out their parents or a trusted adult.[13]

The nature of cyberbullying is important for you to understand. Cyberbullying is

- *24/7.* For the victims, cyberbullying feels like it never stops. They may see posts and threats on their social networking sites. Sometimes the cyberbully taunts the victim at gaming sites and even sends threatening text messages. Bullying used to be confined by the school day, so that victims would get a reprieve at home. Not anymore.
- *Everywhere.* Bullies go online using a variety of technologies and services. They find their victims at home, at school, in the mall, and in bed at night. Because more than 80 percent of teens use a cell phone on a regular basis and never turn it off, this tends to be the way that they are targeted first. In addition to mean voicemails and texts, the bolder bullies take their taunts and threats online, where they can get an audience.
- *Extremely public and viral.* The online form of bullying plays out in front of thousands of people. Recently, there have been some tragic cyberbullying situations in the news that are heartbreaking for all involved. In these situations, bullies didn't handle their anger, sadness, or grievances in a constructive manner. By playing out these emotions

in online attacks, the public shame they caused their victims drove those victims to drastic actions.

We have all read about child suicides attributed to a target's inability to cope with cyberbullies. Here are just two examples of cyberbullying cases that caused damage to the aggressors.

One budding basketball star, Taylor Cummings, was not seeing eye-to-eye with his basketball coaches, so he typed out his frustrations online: "I'm a kill 'em all. I'm a bust this #!@$%^ [expletive removed] up from the inside like nobody's ever done before."[14] School officials took the teen at his word, and they expelled him. With an expulsion on his record, he may never realize his dream of a sports career. Even though his taunts were directed at adults, this situation involves a cyberthreat.

In another case in Seattle, a middle school principal dealing with cyberbullying and physical bullying of a student identified twenty-eight students that were involved and suspended all of them. Suspensions do not sit well with college admissions boards, especially at institutions that have a heightened awareness and sensitivity to cyberbullying.

The Bullying Type

Several states have put into place cyberstalking and cyberbullying laws. If your kid is the bully, there could be legal repercussions that could have an impact on his online reputation.

So who are these cyberbullies? The cyberbully does not have an exact profile, although girls tend to cyberbully more than boys. Otherwise wonderful kids have been known to cyberbully, join in, or quietly watch another kid being victimized. Online bullies might also be physical bullies. Sometimes kids who are bullied at home themselves go online and feel that they have the power and cloak of anonymity to bully others to relieve their frustrations. Other times, kids just join in with other kids because of peer pressure.

Your kid may actually be a cyberbully. If you ever find that your child has bullied another online, after apologizing to the victim, look for ways to clean up your child's online accounts so the victim can no longer see the posts and, hopefully, your kid will not have regrettable lapses in judgment come back to haunt him or her later. Bullies with unchecked behavior rarely grow out of this habit without some emotional scars. Beyond the digital tracks they leave behind, they do not learn good anger management and coping mechanisms for dealing with stressful social interactions. Studies have shown that children who habitually bully others tend to have a higher incidence of depression and sometimes drug and alcohol abuse in their adult lives.

Helping Cyberbullying Victims

In many cases victims are those who are isolated from others. They just seem like somebody who *could* be bullied. Those with higher self-esteem and a large set of friends are much harder to bully. Often it is not the characteristic the bully taunts the victim about (you're fat, Muslim, or wear glasses) but the mere fact that the person is vulnerable that attracts a bully.

Victims, of course, assume it was their weakness that invited a bully to attack, but the reality is that weakness is just a convenient hook for an aggressive act.

Look for certain clues. If a kid seems withdrawn from school friends, avoids going to school, or suddenly avoids the Internet, he or she may be a victim of cyberbullying. Look for behavior that seems uncharacteristic for that person: is she quicker to get angry or get upset? A drop in grades or loss in appetite may be other signs to watch for.

Here are some tips for things you can do to help victims of cyberbullying:

1. *Golden Rule:* Tell your kids that you will not accept any excuses; they are not allowed to post negative posts or taunts about other people online. No exceptions.
2. *Safe Zone:* Instruct your kids to contact you immediately if they are the victims of cyberbullying or if they see it happening to someone else. Promise them a "safe zone," where the two of you will work together on the best solution to notify the appropriate adults. Also, make sure they know you will not cut them off from the Internet, which is an integral part of their lives.
3. *Not Your Fault:* If you are talking to a victim, make sure he or she understands that he or she is not at fault. Often children believe that there is something wrong with them that prompted the bullying, and they need to be reassured that they did not cause the bullying.
4. *Proof:* Your gut instinct might be to delete the hurtful posts, but you may need them as evidence. At a minimum, take screen captures using your computer's Print Screen feature so you can discuss the postings with the appropriate adults or authorities.
5. *Block:* You can block messages from a bully using e-mail or other software filters. Sometimes blocking access to a victim is enough to make a bully stop.
6. *Law Enforcement:* Get officials involved if physical threats, nudity, or fraud are involved.
7. *School Officials:* If the cyberbullying involves classmates, contact your principal and/or your school board. Most schools have instituted zero-tolerance policies for bullying and can help you handle the situation.

8. *Counseling:* If the cyberbullying was frightening or prolonged, seek counseling for your child and yourself.

Digital Is Forever

The fact that kids can expose personal information or get involved in a cyberbullying event is significant because what they do online may last their entire lives. In addition, such exposure may open them up to cybercreeps, pedophiles, and crooks. One way you can help them protect themselves, as you'll see later in this section, is by friending them and helping them understand the importance of passwords.

Kids' Posts Haunt Them

Has your kid posted likes, comments, or pictures of brands that give off a negative image? If your child doesn't drink alcohol or do drugs but does post pictures of movie star substance abusers, it will not come across positively, especially to people who do not know your child.

In another news story, a college student at the University of North Carolina–Charlotte was frustrated after a long, hard day at work waitressing.[15] One couple hung out at a table for three hours and left her a five-dollar tip. She had about one hundred friends on Facebook and tight privacy settings. She needed to vent, so she did what many digital natives do—a quick and short Facebook post: "Thanks for eating at [restaurant], you cheap piece of camper."[16] Her employer was alerted to her post, worried about the restaurant's reputation, and they fired her. She admits she used bad judgment but had no idea it would cost her a job.

Your child should learn this lesson now before online venting leads to a lost job or relationship. Look for negative posts that your child might be creating, from criticizing friends and teachers to self-destructive posts. These posts can be a case for parental involvement but also something to help you teach your child online savvy, before their postings are viewed by others in a critical and negative light.

For example, would you want your child's rants and negative comments to be part of the record for their online reputation when they are being checked out by prospective employers or college admissions reviewers? Both of these sources look for photos, associations, and posts. There are actually cases where prospective students wrote negatively about the college they just visited and then sent in an application for admission. Guess who didn't get accepted?

What might seem like harmless or impulsive fun could not only ruin your kid's reputation but also get them in trouble with the law.

In Washington State, three teens found out the hard way that one quick click on the smartphone can affect their reputation for a long time.[17] The three teens were arrested for sending a naked photo of a fourteen-year-old girl from a cell phone. I wish the story ended there, but it does not. The texted photo went viral and was forwarded to students in four different middle schools in the area. This constitutes child pornography. In situations like this, if these teens are convicted, they will have to register as sex offenders. The girl's humiliation could last a lifetime, and sex offender registration is forever.

Bad Guys Lurking Online

In a story about child predators, Houston, Texas's KHOU Channel 11 noted that Amanda Hinton of the FBI said this about social networking sites and kids: "The predator, basically, it's like a shopping mall. He looks through the social networking sites, he looks for someone he's interested in talking to, and he capitalizes on what he sees their interests were."[18]

By understanding who the predators are and their tactics, you and your kids can stay safer.

Kids Talking to Strangers

A survey by TRUSTe and Lightspeed Research found that 68 percent of young adults have accepted friend invitations from people they do not know.[19] This can create grave repercussions to their online persona because the strangers may entice kids to do things online that their true friends would not. In the same survey, eight out of ten parents wished they could have more control over what their tweens, teens, and young adults are posting online, including a delete function.

Internet studies have shown that kids can tell you they know not to talk to strangers online and they know not to give out their personal information. However, if you put a pop-up window or ad in front of them promising prizes and money, the cautionary part of their brain turns off and they will click and tell everything for a chance to win the prize.

This is especially true of younger children. During Internet safety classes with kids in grades K–5, many admit that they know they should not fill out survey forms or click on the jumping frog on the screen, but these temptations are simply too tempting to pass up. Besides, what if that sweepstakes entry or pop-up game was the real deal that one time? Often we hear that, after clicking on pop-ups, kids' family computers froze and the families had to pay to have them fixed. This is not just annoying but a potential danger to your entire family's online personas. If your kid introduces a virus onto the home computer, the computer virus may be used to collect personal information about your family members that could be used to commit identity theft.

Ask your child if he or she would leave the front door of your house wide open all night long for any criminal to enter and abuse everybody in the family.

Kids may seem rebellious at times, but no matter how rebellious, once they understand this type of analogy, they get the need for safety. Nobody likes to get scammed or robbed.

Bad Guy Tricks

An eighteen-year-old man in Wisconsin posed as a girl on Facebook to trick young men into sending nude photos. He was sentenced to fifteen years in prison.[20]

In another sad story, three teenage girls were at a party, visited an online chat room, and flashed their breasts. The party ended, and the girls went home. Within a week, one of the girls started getting threatening e-mails. The threat? That pictures were taken when she flashed her breasts and the creep would post them online to all her Myspace friends unless she provided more sexually explicit photos and videos. Instead of telling an adult, she complied with the perpetrator's disgusting requests. Eventually, she asked for help. Police and federal authorities became involved and found the cybercreep. The nineteen-year-old man was charged with sexual exploitation.[21] Think this is rare? Think again. This practice even has its own term, *sextortion*.

Recently, one of the authors of this book, Theresa, reconnected with friends from her college years and colleagues from earlier in her career via Facebook. It was so exciting to catch up, see photos, and see how much their families have grown. The more she clicked on the profiles, the more she learned about the kids. Thankfully, all the kids had a positive online image. Theresa did not see any heavy partying or negative posts. However, if she had been a cybercreep, they had all left enough digital tracks and clues online that she could have easily impersonated the kids or contacted these kids and impersonated a friend of their parents.

In one situation, she was able to see photos of the kids getting ready for a prom, learning the kids' names, the places they visited, and the address of their homes. Online there was a full and open chronology of the events in these kids' and young adults' lives. The parents' privacy settings were moderate, but the kids' settings were wide open.

A kid on a laptop in a bedroom is exactly what child predators are looking for. In a child pornography arrest, Jason Bezzo was accused of having a large child pornography collection. He had seven hundred gigabytes of child porn—that is, hundreds of thousands of photos and videos. Authorities believe that he obtained a considerable amount of his collection while he was visiting the kid-friendly video chat sites blogTV and Tinychat.[22]

It is critical that you protect your kid's online persona from snooping cybercreeps and cyberpredators. In the next section, we tell you how.

Raising Good Digital Citizens

There are several things you can do as a parent or caregiver to help kids to not only stay safer online but also put forth a more positive online identity that will serve them better in later life. As a side benefit, you can help to ensure that nothing your child does online will reflect badly on you. Whether your son or daughter is e-mailing, chatting, texting, gaming, or connecting through social networking sites, the information in this section will help you keep them on track.

First of all, it is important that you talk to your kids about their digital persona as early as possible to avoid problems later in life. Before they have an e-mail or Facebook account is the right time to set ground rules and have conversations. If they already have one, plan to begin talking to them about their digital persona now. You may read this and shrug your shoulders, saying, "My kid is a good kid and has a good head on his shoulders." That is probably true, but the Internet is still a test bed for them as they learn the ins and outs of having freedom online and growing up.

Although most of the big social media sites do have age restrictions, anyone with a little tech savvy would be able to set up an account, and it is really hard for these sites to definitively prove the person signing up is the age they say they are. The father of a young girl recently sued a major social media company for failing to enforce its age restriction policy after claiming his daughter was exposed to sexual predators when she signed up for an account at age eleven. Granted, she seemingly violated the site's age policy but parents can use this as a teachable moment.

Discuss your expectations with your younger child. Tell them you don't want them to get a social media account without discussing it first. If you find that your underage child has created a Facebook or other social media account, contact the site and ask them to delete the account promptly. When you're ready to allow your kids to access social media, make sure they don't share their personal details (date of birth, school, and location) on any social media site.

Follow Them There

You need to be where your kids are online. In the United Kingdom, a study by the National Citizen Service asked kids where they seek comfort when they are stressed. The girls in the study said they seek advice or comfort on social media before they talk to their parents.[23]

Teens and millennials in the United Kingdom are using social networks more often than any other age group, using apps such as Snapchat and Instagram to stay in touch.[24]

In the United States, Facebook, although not exclusively used by teens, is still a popular social media landing spot for kids ages thirteen to seventeen. Seventy-one percent of teens are using the site, and half of teens say they also use Instagram and 40 percent like to use Snapchat.[25]

There is also the popular After School app. What we do not like about this app is that it breaks our cardinal rule of telling you where your kids are. No adults are allowed as members. What we do appreciate about this app is that they have built in a few safety features that parents and teachers alike can appreciate. It has an algorithm built in that looks for keywords and can trigger a fast-response system that contacts authorities if a threat is detected. If needed, police can trace a post to a particular device. The app is also equipped with a warning system that flags any worrisome messages about being depressed or angry and triggers a message asking if they would like to text with a counselor. If parents know about the app, they can set filters to block certain content.

E-mail Guidelines

Rites of passage for your kids include following the latest fashions, listening to new music, and getting their first driver's license. These are all experiences you enjoyed as a tween and teen. But there are things you did not experience— for example, getting your first e-mail or social media account and posting your first information online.

Rules for determining the right age for e-mail accounts, social networking accounts, smartphones, and other access to the Web are not hard and fast. However, if you need someone to play the tough guy when you tell your kid no, blame the online companies' age guidelines. Several Internet e-mail service and social network providers do have age limits and require kids to be at least twelve to thirteen years old to access their services.

However, if a service has no age limitations, age should not be the only factor in your decision. Another factor to consider is the maturity level of your children when they are happy, sad, or mad. If they are still immature or over-emotional, Internet e-mail and social networks might not be a good fit for them.

If you decide to let your child have his or her own e-mail account, consider these areas of vulnerability to keep your kids and their online personas safe.

- Account Name: Choose an e-mail address that does not identify a child's name, age, or gender.
- Rules: Discuss ground rules about appropriate e-mail communication.
- Let them know that they are not really anonymous and that cyberbullying, sexting, and sending pictures via e-mail are dangerous. Make it a rule to avoid clicking on links in e-mails. Tell them that the rule of "don't talk to strangers" and the Golden Rule both apply online.

- Attachments: Tell children not to open attachments without consulting you first. Kids are notorious for clicking on all kinds of links and sending infected attachments to others.
- Review: Tell them you will be reviewing their e-mails regularly, and ask them to make sure their friends know that their e-mail account will be monitored.

Online Chats

Chatting is one area where parents feel defeated. There are free services popping up periodically that make it easy for kids to exchange text and video chat with friends and strangers. Many parents block these sites but, after a while, get overwhelmed trying to keep up with new ones. Like Lucy and Ethel desperately trying to keep pace with the chocolate factory production line, their good intentions can't keep pace with the number of chat sites out there.

In conducting research for this book, we visited several video chat sites. There were three reoccurring themes across the sites that we noted.

- Almost every young person online was in a bedroom while chatting.
- More than 50 percent of the kids were sitting on their bed conducting video chats.
- There were no parents in the background monitoring their chats.

Discuss house rules for technology use when your kids are at school, home, or away from home. When kids are in their bedroom on a laptop or smartphone with video chat capability, this is a recipe for disaster. Kids connected to the Internet should be in a common family space, even if you are not looking over their shoulders.

Make sure your kids know that chat rooms are wide open, easy to use, and although some harmless fun can be had there, these sites are a magnet for pedophiles and other creeps.

In addition to children's video chat sites, there are online video chat services that were designed for businesses and consumers that are used by kids. The good news is that many of these offer privacy settings, so you can lock your kid's profile down and keep strangers out. These services include: Skype, Apple FaceTime, AOL's AIM chat service, and Yahoo! Messenger.

Some sites you may want to visit to do your own research are listed here:

- www.ChatRoulette.com
- www.Tinychat.com
- www.blogTV.com

Be Aware of Secret Apps

One category of apps is a growing concern for parents who do their best to keep tabs on their teens' digital activities on smartphone and tablets. These so called secret apps may look like your average smartphone app, but there's a huge market for secret apps, also called *decoy apps,* that can hide activity on a cell phone.

There are often legitimate reasons why a person may want to hide messages on their cell phones. They might be protecting sensitive, personal, or work-related information or safeguarding against phone theft. Your kids may be testing the boundaries of their individual privacy, or they may be using the apps with less pure intentions.

The names of the apps will change over time but here is a short sample list of apps as of the writing of this book that you may want to look for.

- 9Gag allows you to send memes and pictures online. The content isn't currently moderated and there have been past reports that 9Gag has been used by cyberbullies. Some swatting cases (when someone calls 911 and convinces police to raid an innocent person's house) are alleged to have originated on 9Gag.
- Cover Me hides text messages, documents, photos, phone calls, and more.
- Secret Apps hides texts, photos, and more and will secretly take a picture of anyone who tries to access your files. Secret Apps looks just like a normal Apple iOS app.
- Vault-Hide hides message and data in a vault. This vault is password protected and camouflaged.
- Hide SMS ensures that incoming SMS messages don't show up on the phone and go into a vault. The vault notifies the user when messages arrive.
- Hide Text SMS & Calls hides your call logs, photos, and texts.

Want to know how big this market is? Go to your app store and type "secret" in the search bar. You'll be amazed at the apps you find and the number of times they have been downloaded!

Show, Don't Tell

The best rule of all for parents is "show, don't tell." Parents should use social networks and set the example. A great way to share privacy and security concerns with each other is to talk with each other. Share news headlines that show both positive and negative stories about social media. You can also play Internet safety games together to make sure that everyone is up to speed on the latest threats. Some of my favorites are at NSTeens.org and OnGuardOnline.gov.

Don't make assumptions about the safety of any app. Carnegie Mellon conducted a study in 2014 that looked at more than one million apps to evaluate user privacy. They found that some of the fun games like Fruit Ninja or Angry Birds were some of the bigger offenders of collecting user information, whereas Facebook and some Google apps actually did a better job guarding privacy.

A great way to really understand what the privacy policy of any site or app means is on CATSMI.CA. You can also try out the new Privacy Grade database at PrivacyGrade.org for more information on Internet privacy policies.

A Primer on Texting

Love it or hate it, it's time you became a lot savvier about texting. Nielsen surveyed three thousand U.S. teenagers between thirteen and seventeen years old. Girls sent roughly 4,050 text messages per month. Boys sent roughly 2,539. You don't have to be a math whiz to determine that that's a lot of text messages per day.[26]

In a 2010 survey, 42 percent of teens said they can text with their eyes closed. Some 43 percent of teenagers say texting, not safety, is the number-one reason for asking for a cell phone. About 22 percent of teens prefer texting over phone calls because they consider it easier and faster.

Do you feel as if you need a translator to come with you when you read your kids' posts? You are not alone. Many of the posts are a combination of strange abbreviations, hashtags, and graphical symbols called *emojis*. The strange abbreviations might be something called *leetspeak*. Hashtags are usually words jammed together with the pound sign, the "#" in front of them for emphasis. Emojis are those smiley faces and other characters you might see in your text messages. We are all aware that to keep up with the times, we have to keep up with things such as new TV shows, Top 40 music, and the latest slang used by our kids. In the case of keeping up with emoji, hashtags, and leetspeak, the shortened messages that your kids use when they text on their smartphones, understanding what kids are saying can help you keep them safer.

Emoji Revealed

If you are not using emojis when you text with someone under the age of twenty, you need to read this section to get up to speed. Emojis tend to have different meanings around the world. In a recent study of emojis tracked by the company Swiftkey, they found that Americans used the most random emojis and often used the cake emoji more than other countries. Canadians tend to use emojis focused on money, sports, and violence. The French and Russians use emojis with hearts more often than other countries.[27]

Want a translation to understand and use emoji properly? Ask your kids; they are the best and most current source. You can also visit http://www.iemoji

.com/popular/emoji. This site tells you each emoji's history, if it's popular or not, and many of its common meanings.

When we asked kids why they liked to use emojis they said they found it easier to convey their true feelings. When delivering a tough message if they added a heart or a smiley face to it, that means they care about the person enough to deliver the message. Sometimes, kids said that emojis told everything without having to type a word. This confirms the old adage, a picture is worth a thousand words.

Leetspeak Speaks for Them

Kids are so good and quick at communicating on their phones that leetspeak is becoming a part of our language. Here's a list of ten text message leetspeak terms that kids commonly use. See if you need to brush up on your leetspeak.

IMHO (in my humble opinion)
PAW, PIR or 9 (parents are watching, parents in room)
TBH (to be honest)
IDK (I don't know)
Totes (totally)
BFF (best friends forever)
<3 (heart)
2H2H (too hot to handle)
420 (let's party—could also involve drugs/alcohol)
,!!!! (talk to the hand)
53X (sex)

Did you know what all of these meant without looking them up? If you didn't, you're not alone. If you would like a handy source, try asking your kids to tell you what the leetspeak means or visit this reference guide at: http://www.abbreviations.com/acronyms/SMS.

As of this writing, there are several translators and guides to understanding text messages. You can use these free tools if you must, but the best way to learn and connect with your kids is to ask them what a text message means. If you're not sure you believe the answer, then look the term up by typing the text message into your favorite search engine.

Sexting 101

You might not want to think about it, but many teens are sexting. A 2012 study in the *Journal of Pediatrics* indicates that sexting is a new norm among young adults. The National Campaign to Prevent Teen Pregnancy[28] shows that nearly 40 percent of all teens have posted or sent sexually suggestive messages.

This may make you uncomfortable but talking about sexting is probably the most important talk you must have with your tween or teen. Here are some tips for how to approach the topic:

1. *Establish a Safe Zone.* Teach your kids to talk to you about these issues and promise them that you will not overreact.
2. *Be Helpful.* Think about what you would do if your child did send a nude photo to someone and it is now making the rounds. Focus on helping, not punishing.
3. *Rehearse.* Someone will ask them, at some point, to text a nude. Work through some options for how they want to handle that situation, or offer a few canned responses. You might use humor to suggest that your kid consider sending nude animals such as chicks, bunnies, or puppies. Use the morality line: my parents won't let me do that. You can also suggest they simply ignore the request and tell the requestor in person that you are flattered but you don't do these.
4. *Be Alert for Sextortion.* Sextortion is real. It hurts. Often laws and law enforcement cannot offer much help to the victims.
5. *Legal Issues.* The distribution of nude photos of someone younger than eighteen is illegal. The parties involved could be charged with exploitation of a minor and possession of pornography. This can result in fines, jail time, and having to be registered forever as a sex offender.

Keeping Tabs with Location Software

Most of us remember growing up with a favorite hangout where we could connect with our friends. With many kids going to schools across town and taking part in far-flung after-school activities, making such connections is not as easy as it once was.

Today kids check in with friends and let people know how to find them by using check-in or location software. Some popular location services used today include Facebook Places, Waze, and Foursquare. These services are free and encourage frequent check-ins. Foursquare check-ins can result in a person being named "mayor" for the most visits or even presented with coupons and discounts. This is a great and fun way for kids and adults to stay connected to their friends. However, from a security standpoint, your kids are clearly broadcasting where they are and where they are not (e.g., at home). Would you want someone scrutinizing where your kid hangs out and judging him negatively based on where he or she spends his or her time? What is the perception of a kid who is the mayor of the local game place or mayor of a convenience store in a seedy part of town?

Talk to your kids about these services. If they are using them, you should, too. Test them out and make sure your family is aware of the trail of information about their activities they are leaving for all to see.

If your main concern about your teen's location is his or her driving safety and phone use while driving, there's an app for that called Canary-Teen Safety. This free app is designed to stop distracted driving by sending you notifications in real-time when your child is engaging in risky behavior. For example, the app lets you know if your child is using the phone while driving, exceeding a speed limit that you set, traveling into areas that are off-limits, staying out past curfew, or traveling near possible bad weather.

Read more about this app on FamilyEducation.com at http://fun.family education.com/mobile-apps/internet-safety/75744.html.

Your Kids and Online Gaming

Kids love to play games. In a recent Internet safety class, we were talking to kids in kindergarten and first grade. We were going over the rules with them and getting lots of nods, but the kids were not enthusiastic until Theresa asked, "By a show of hands, who likes to play games on the computer, Internet, game station, or smartphone?" All the hands went up, and several kids said, "You forgot to mention the iPad!" When we asked these younger kids where they game, they said they are usually in a public area of their houses.

This answer does not hold true for middle school kids, who will play a game anywhere. Most of them say they go into a bedroom or family room so the gaming noise doesn't disturb their parents. In the spirit of "be where your kids are," this is the wrong answer.

Many electronic games can link two players or two thousand players from different locations to chat and play together.

Here are some tips for protecting your kids when they're gaming online.

- Remind your kids that everyone is a stranger. If a friend invites him or her to game online, have him or her call that friend first to verify that it's really that friend.
- Make sure your home computers are up to date with the latest versions of firewall software and antivirus and antispyware.
- Create a safe zone. Make sure your kids know that you will not take away privileges or stop them from playing if they come and share a situation with you that makes them uncomfortable.
- Take time to play. Prescreen all your kids' games before they are allowed to play them. Even if you are not a great gamer, you can play games set to "demo" or "easy" to get the idea of the game and how it works. Ask your kids to play their games with you once in a while.

- Account names are important. Just as an e-mail account name should never be your child's nickname or identify him or her as a young boy or young girl, a gaming account name should not give away personal information.
- Choose safe pictures. Some games encourage you to post a picture with your account name. Instead of posting a picture of your child, have him find a fun avatar or picture of a cartoon character to use.
- Block text and voice chat on games.

Your Kids and an Online Moral Compass

There's no question that teens are far more technology-savvy than their parents in most cases. The question is whether they know when to use the tech skills they've acquired and when to stop.

We see it in the headlines. The hackers who broke into the UK cell phone company, TalkTalk, in 2015 were younger than age nineteen. The hackers who broke into the personal e-mail accounts of the U.S. Central Intelligence Agency Director and Director of the Department of Homeland Security in 2015 were also younger than nineteen. There was a case in the United Kingdom where six teens between the ages of fifteen and eighteen were arrested for using a tool called LizardStresser, an online tool for attacking websites. Authorities say they hacked into a newspaper, their school, and gaming companies. Teens were swept up in what police called "Operation Vivarium." This was led by the UK's National Crime Agency (NCA). The NCA was watching the Lizard Squad that offered the tool online and which is the same group suspected of attacking Sony and Microsoft on Christmas Day of 2014. The teens suspected allegedly purchased the hacking tool using alternative payment methods, including bitcoin, in an attempt to fly under the authorities' radar.

You must have the moral compass talk with your kids. Just because they can doesn't mean they should and rules in real life also apply online.

Your Kid's Persona on Social Networks

Social networking can be a great way for kids to stay in touch and share experiences with others. However, it's important that they understand the risks associated with social networks. If they share their pages with too many friends, eventually they will share their information with a stranger.

In this section we discuss the dynamics of friending online and provide overviews of various social networks that are available today.

What is the best age to start using social media? This is a question parents struggle to answer. The age at which a child starts using sites like Facebook,

Twitter, Snapchat, and others really depends. A starting point is to determine the site's age requirements.

Your child should at least meet the minimum age guidelines of each site. Read the terms of service for each site he wants to use to determine the minimum age.

Here are some sample site policies. Twitter's privacy policy, states "Our Services are not directed to persons under 13." According the policies on each of their websites, for Instagram, Facebook, Pinterest, and Snapchat the minimum age is 13. LinkedIn's minimum age is 14. YouTube is 18, but kids aged 13–17 can sign up with a parent's permission.

Live Streaming Apps

Taking live pictures on the go is easier than ever. Smartphone apps such as Periscope and Meerkat allow you to send a live feed from wherever you are to followers all over the world. They're cool, immediate, and it's easy to get addicted.

When it comes to kids, their Internet identity, and online brand, there are some things to think about before they use these services.

What Is Live Streaming?

Live streaming allows you to videotape from your smartphone, tablet, or laptop and have a live World Wide Web audience watching without any time delays. Facebook added live streaming capabilities for all of its users in 2016.

Two popular apps are Meerkat and Periscope. They both allow rapid video filming and streaming in real time and you can easily publish videos to social media sites.

What Are the Internet Identity Concerns?

You should be concerned if a streaming app has:

- Location tagging. If they know where you are and where you have been, hackers can predict where you will be.
- Unmonitored comments. Both apps mentioned above monitor content to an extent, but users have said that comments appear to be unmonitored. This is like dropping your teen off at a rated R movie that you didn't watch first.
- Claims of sexual harassment. Women have reported participating in live stream conversations only to have the viewers say lewd and suggestive things while the live stream was underway.

As with many new platforms, it takes a while for operations centers to figure out how to block fraudsters, predators, bullies, and anyone else you don't want in contact with your children. A side legal concern for your kids is that, in some states, videotaping or recording without permission in a private space can be considered a crime.

Here are some live streaming tips for you and your kids:

- Talk to your kids and lay down some ground rules about live streaming.
- Try it out as a family so you know how it works.
- Turn geocodes for location tracking and pictures to off.
- If someone is bothering your kids on a live streaming platform report it to the hosting site right away. For example: On Meerkat you can type in #911MK + user name.
- Look at your surroundings. What could be used to track your kid's identity in a live stream?
- Clean up Twitter accounts. Are there geocoded photos, pictures of your house, or other information useful for tracking your child?
- If, after using live streaming, you fear for your kid's safety, seek help from law enforcement. Victims can file a report at IC3.gov, the site of the FBI Internet crime desk.

Do You Know Your Kid's Online Friends?

Parents are used to meeting other parents before dropping their kid off at a new friend's house. When your kids go out somewhere, you ask who else is going to be there, how long they will be at each place, and you may even ask them to call you when they get there so you know they are safe. With social networking, your kids may be sitting at home, but they are going into virtual worlds. Do you know the kids they chat with? Do you know who else is going to be there? Not knowing who your kids are connecting with online is just one problem.

Another is how you try to connect with them. Many parents use e-mail, but that is not the communication tool of choice for kids. Kids are texting and flocking to Facebook. Some parents who are on Facebook know that you can set up your kids' accounts to send you an e-mail when they post content or a note is posted to their walls. However, most do not realize that kids can hold an instant chat on Facebook and the service doesn't send an e-mail or leave a record online. Because Facebook makes it easy to video chat from their service, anyone from friends to strangers can have visual access to your kids without leaving an obvious record. In a speech, retired general Colin Powell shared a personal story. He talked about how hard it is to stay current. He used an example of trying to stay in touch with his grandkids, and he asked them why they never respond to his e-mails. His grandkids told him, "Poppy, nobody e-mails anymore. You need to use Facebook!" Then they set him up with a Facebook page. If we do not connect to the kids where they are, we might as well be in different countries or speaking another language. The fact is that parents are blogging and adopting social services like Twitter, but kids aren't. At some point, parents and kids are tethered to the Internet, but not in the same way, so they completely miss each other.

What is the answer? Well, it's definitely not the stance of one principal at a New Jersey middle school, who feels so strongly about the potential downsides of social networking he sent an e-mail blast to parents asking them to consider taking down their kids' online social networking profiles, including Facebook.[29] Anthony Orsini, principal at Benjamin Franklin Middle School, also wrote, "There is absolutely no reason for any middle school student to be a part of a social networking site! Let me repeat that—there is absolutely, positively no reason for any middle school student to be a part of a social networking site!"[30] Not only is it virtually impossible to keep kids from the social Web, but it's also not productive, as eventually they will have to learn how to be safe, productive online citizens. Start educating them now about safety, and stay in touch with what they're doing online and with whom they're doing it.

Keeping up with what your kids are doing online doesn't have to feel like a losing battle. Review your kid's profile. It's important that you friend your kids and be where they are online. Some 70 percent of parents in one report said they don't know what their kids look at or read while they are online.[31] If you don't know what your kids do online, you can't guide them in making good decisions that build good online reputations and protect them from cybersnoops or creeps.

GANGS ONLINE

Did you know that social networking via the Internet is the new recruiting tool for gangs like MS-13? It could be damaging to your child's online reputation, not to mention his or her safety, if he or she is linked to gangs. Young people who are surfing online may come across pictures, music, or videos about gangs. Of course, children are curious, so they click and see something that glorifies gang life. That's when a gang member may strike up a chat with them online and try to lure them into joining their club or linking to their social networking page. About 70 percent of gang members say it is easier to make friends online than approaching kids in person.

Take Inventory of Social Networking Sites

So, just where might you find your kids online? When it comes to social networking sites, Facebook is much touted, but it's not the only one out there. There may be one that fits your or your kids' needs better than others. Here are some of the most popular sites used by U.S. teens and teens around the globe[32]:

- Facebook: The most popular social network that also has a location check-in service called Facebook Places. Ages thirteen to seventeen make up 10 percent of Facebook users. The largest age segment on Facebook is eighteen- to twenty-five-year-olds (29 percent). Some 61 percent of Facebook's users are thirty-five or older.

- Instagram: Picture posting site allows you to add comments as well. You can share, like, or comment on other posts. One-third of teens responding to a survey conducted by Piper Jaffray said it's the most important app they use. Instagram is owned by Facebook.

- Snapchat: A video messaging app that allows you to send a video to your friends. The video lasts between one and ten seconds and is removed from view after the recipient views it. The video will eventually be deleted from the Snapchat servers but as we have said before, once you post something, it's not yours and you may not be able to delete it. The recipient can videotape the video as it plays and keep his own copy to replay or distribute later.

- Twitter: A microblogging community where members can send public updates and links to sites using 140 characters or less. Twitter is not as popular with younger kids as with young adults; Twitter also offers a direct messaging feature where you can carry on a conversation with another member in chunks of 140 characters or less. All tweets are archived by the Library of Congress, so be careful what you include in public tweets, as they will be captured forever.

Young Kids: Are There Any Safe Havens?

OnGuardOnline.gov ran an article that noted that social networking sites are attracting preteens and even kids as young as five years old.[33] Providing a safe zone where young kids can put positive online behaviors to practice is a good idea.

There are many social networking sites that have been designed to provide parents with peace of mind and to protect kids. Older kids will probably still want to join Facebook or Myspace, but you have options for your younger ones that provide a safer environment.

Some options include the following:

- Yoursphere.com: Designed for seventeen-year-olds and younger, the site offers social networking and features for kids that range from sending messages to gaming. There is a portion of the site dedicated to giving parents tips and the latest information on kids and Internet safety.
- Togetherville.com: Parents can use their Facebook accounts to create a profile for their kids on this site. The target age group is ten and younger. Parents can view and post messages to their kid's wall. Each member is asked to agree to a code of conduct covering behavior such as cyberbullying or posting negative messages online.
- Kidzworld.com: Kids can connect with people they don't know but the site has both software and adult monitors and strict privacy and safety rules.
- helpyourhero.site-ym.com: The premise behind the social network is that it transforms kids into superheroes. Kids can connect with other family and friends that you identify.

Parents: Your Next Step

To get started keeping your kids safer online right away, here are a few things you can do.

- Set up your own profiles on some of the more popular social networking sites and make a commitment to yourself to use them, at least once a week. Many of the social networking sites now offer smartphone apps that make it easy for you to click an icon and see the latest posts, even on the run.
- Teach your kids that online posts should be supportive, not destructive. In the physical world, we encourage our kids to be positive and to be friends with other people who are positive. We tell them to support their friends when they are down. We encourage them to cheer for their teammates, even if they are losing or someone makes a mistake. These life lessons in the physical world are even more important in the online, virtual world. Even one story of deep depression or suicide due to online posts is one story too many. Parents must take steps, early and often, to avoid hateful and destructive posts that hurt children deeply and make them feel that the whole world thinks they are worthless.
- Help your child build an online network that is connected to sites that are positive. Help them connect to people who are positive. This is the best way to build a supportive environment around your child.
- Many parents feel that because their kids are not making purchases or conducting online banking that they are not targets. That's not

true. Their passwords on nonfinancial accounts safeguard them from identity theft (more about this shortly) and harassment. Your kids need to know how to create strong passwords, how to protect their passwords, and how to have overall good password maintenance.

Password Woes

Here's the story of one child who became a victim after sharing his password with his best friend. The best friend decided one day he would log into that account using the password his friend supplied. Once into the e-mail account, the intruder sent out mean and hurtful e-mails from that account. The child, whose account had been hijacked, was dumbfounded when his parents told him he was in trouble and showed him e-mails sent from his account. After finally convincing his parents he was a "victim," Theresa was called in to help them determine who hacked into the e-mail account, what recourse they had with the e-mail provider, and how to get the bad guy out of the e-mail account. In the midst of tracking down e-mail headers and IP addresses, the victim remembered he had given out his password to another account to his best friend. Because he used the same password on all his accounts, the best friend had free access to his virtual life.

- An important reason to not share passwords that give access to personal information online is the growing trend in kid identity theft. It's shocking but true; cybercriminals are targeting kids. Cybercriminals buy or generate lists of dormant Social Security numbers. When they find a number that has been assigned to a young child, it's like hitting pay dirt.
- Cybercriminals love kids' Social Security numbers because they are rarely tracked. They sell your kid's number and help people run up credit card debts they never plan on paying. In some cases, once they realize they have a kid's Social Security number, they may surf social networking sites looking for clues and facts to make their stolen profile a little more complete and more enticing to use and sell to other cybercriminals who want to commit fraud. Often, families don't know this is happening until their child applies for a student loan or opens a new bank account and they find out the child's financial reputation is ruined.
- A key warning sign that you may have an issue in this regard is if you get credit card applications in the mail addressed to your child; this could be a sign that a cybercreep has started a credit history in your child's name.

PROTECT YOUR CHILD'S CREDIT

Here are some important steps to protecting your child's credit:

1. When you open bank accounts for your kids, ask to have their names removed from marketing lists.
2. Request a free credit report from Annualcreditreport.com every year and teach your child how to do so once he or she is old enough.
3. If you are worried that your child's Social Security number has been compromised, you can request a credit freeze on credit reports until the child comes of age. By freezing access to a credit report, you stop just about any bank or store out there from approving an application for credit. When your child is ready to apply for credit, he or she can lift the freeze temporarily.

Just-in-Time Parenting

Online parenting is not something you do once; it requires setting up the ground rules and then instilling daily habits for yourself. You need to be where your kids are online so you can be just in time to help them avoid potential problems or reward and encourage good behaviors.

When you are where your kids are online, teach by example. Show your kids the positive and fun side of being online. Your posts should show support for your friends and family. Make sure you only post notes that are respectful and don't reveal personal information. If you have grievances with a person or company, show your kids how to handle them in a positive manner. By practicing just-in-time online parenting, you may be able to head off potential problems before they damage your kid's reputation or future.

Trust is important, but you still need to verify that your kids are following the family rules and protecting themselves and their reputations online. A great way to know what your kid is saying online or what other people are saying about your kid is to set up a Google Alert to be notified whenever your kid's name appears online.

To set up a Google Alert, follow these steps:

1. Go to www.google.com/alerts.
2. In the search terms box, type "your child's name" within quotation marks.

3. Select the type of alerts as "Comprehensive."
4. Select "How Often." If your child is active online, "once a day" or "as it happens" may be the best choice.
5. Type in your e-mail address.
6. Click on the "Create Alert" button.
7. Go to your e-mail inbox and click on the link in the e-mail from Google Alert to activate your alerts.

The Risks of Not Showing Up Online

Here's one example of a mother who wished she'd practiced just-in-time parenting. If Marie and her ninth-grade son had only known that they should monitor his name, they might have avoided a negative situation.[34] After Marie noticed her son was withdrawn, she kept prodding him and asking what was wrong. He finally told her that the kids at school were upset with him for all the nasty posts he made about them on Facebook. There was only one problem— Marie's son did not use Facebook and he did not have a Facebook account. Mother and son went online and were astonished to find a Facebook page for the son that included his name and a picture of him. His wall was full of nasty posts, many about people her son did not even know.

Someone had taken on her son's identity and was posting the nasty messages. Marie went to the police and went through a long, arduous process to find out who was behind the nasty Facebook smears using her son's persona. The police had to subpoena Facebook for the computer network address or IP address. Once the police had that, they had to subpoena the Internet service provider to get the home address of the computer's owner. The police finally found the culprits, three young men, one of whom had been a friend of the victim since preschool.[35]

Your Kids Are Naked—Who's Watching?

Often parents are focused on the bad guys tracking their children and don't realize that there is a whole network of people reviewing, checking on, and judging your kids based on what they post online. Some of these people are snooping and trying to take advantage of your children. Others have a legitimate reason that is relationship based—whether checking them out as a suitable babysitter, reviewing their application to college, or accepting them to the cheerleading squad or football team.

People are also turning to the Internet to search for information before dating someone, including parents who may be checking out your kid to see if this is someone they want their child to date.

What Behaviors Are Others Looking For?

Some mistakes that kids make in the digital world can have an impact on their online reputation and follow them for years. Areas to watch for when you monitor your kid's accounts include those listed here:

1. TMI: Sharing too much personal and private information.
2. Bullying: Do they bully or disparage others openly online?
3. Laying Down with Dogs: Too many "friends" makes it hard to manage a social network and in turn, their online social profile.
4. Mirror, Mirror: Personal videos/pictures that do not show them in a positive light.
5. Anonymous: Using anonymity to pretend to be someone else and using that fake persona to post negative content or engage in negative behaviors.

Consider People They Need to Impress

Building an online persona is done in part to protect and in part to impress. There are many people your kids will need to impress as they move through school and into their adult lives.

Fully 70 percent of job recruiters questioned in a survey indicated that they have rejected candidates as a result of information they found about the person by searching online.

According to Microsoft's survey, "Online Reputation in a Connected World,"[36] recruiters are scouring several sites. It's important that your kids know where they look and compare it to where kids like to spend their time. Some fascinating statistics from the survey included the following:

- 27 percent of the recruiters interviewed for the Microsoft study go to online gaming sites as one of their sources for checking out an applicant.
- 63 percent go to social networking sites.
- 59 percent go to photo- and video-sharing sites.

These are the places kids like to hang out online, so it's important that they know that somebody's watching, other than their parents.

In a telephone interview conducted by Kaplan, 31 percent of the college admissions staff taking the survey said they checked applicants on social media sites and 30 percent of the college admissions staff noted that some social media posts negatively impacted an applicant's prospect of being accepted into the college or university.[37]

Studies indicate that at least 25 percent of colleges are using search engines and social media to review the applications of prospective students. Rules and policies for college admissions vary, and not all colleges are sold on this idea. According to interviews conducted by the *Wall Street Journal*,[38] there are still many college admissions offices that shy away from using search engines and social networking sites to consider declining an application.

The State University of New York at Binghamton, for example, sees social networking for kids and young adults as casual, unofficial conversations. Sandra Starke, the vice provost for enrollment management, said, "At this age, the students are still experimenting. It's a time for them to learn. It's important for them to grow. We need to be careful how we might use Facebook."[39]

S. Craig Watkins, associate professor of radio, TV, and film at the University of Texas, was interviewed by the *Chronicle of Higher Education*.[40] They asked if he thought college admissions should research prospective students on the Internet. He felt it was okay, as long as the admissions office did not use this approach to intentionally dig up information that was "gotcha" in nature. He also said, "It is an opportunity to learn about people's interests, the kinds of things they are engaged in, in terms of community-related issues and social issues. In that sense, it does provide a window into a person's life, and into a person's interests that can be a value to an admissions committee."

The *Wall Street Journal* talked to college admissions counselors about how the Internet is used during the applicant screening process. Janet Rapeleye, the dean of admissions at Princeton University, spoke up, and her input is invaluable for all parents with kids seeking to apply to college: "I think students have to expect that if there's anything public, it's possible that we might see it. If there is something that is compromising on your Facebook page, or that you have done on the Web that you may be not proud of, you should probably do everything you can to get that cleaned up before you get into the admissions process."[41]

Even though not all schools have bought into this type of research yet, if you have a child in middle school or high school, you should just assume that by the time he or she applies for college admission, this will have become a routine part of the admissions process.

Helping Them Dress Their Internet Persona for Success

It's time to start helping your kids build a positive persona that will help them throughout their lives.

Pay for It

There are paid services you can use to help you monitor your kid's activity online, whether on a computer or smartphone. Some examples are:

- Internet Monitoring: Software that tracks your child's e-mail address and monitors posts. Services vary, but many of the monitoring software

KNOW YOUR RIGHTS

You do have some support when it comes to younger children that will help protect their online persona. The Children's Online Privacy Protection Act (COPPA) is an important piece of legislation. This is a federal law that requires that any Web applications targeting children thirteen years of age and younger have parental permission before the website can collect personal information. They also have to have parental permission before they can share or use that collected information. It is not a foolproof tool, but it has kept some companies from being too aggressive in enticing your kids to give up information.

There are also cyberbullying and cyberstalking laws in effect at both the state and federal level.

offers tracking of one or more social networking sites such as Facebook, Instagram, and Twitter activity. Some examples of tools that do this are SafetyWeb and SocialShield.

- Cell Phone Monitoring: This service captures all incoming and outgoing text messages and phone numbers and offers you the ability to disable text while driving. GPS technology can help you to locate where your kid's phone is. There are many tools you can buy that offer various features, including KidPhone Advocate and CellSafety.

Having Fun Building a Positive Persona

Although there are a lot of warnings in this chapter, kids and their parents can have fun building a positive persona online. Talk to your kids about their dreams. Discuss what they want to be when they grow up. Talk about the steps they would need to take to achieve their dreams. Create a plan for mapping out the person they want to grow up to be and how they could create a virtual representation of that person on the Internet.

There are some personal maintenance steps you might want to take to keep tabs on and control your child's online reputation, including the following:

- Search: Sit down with your kids and use search engines such as Google and Bing to find out what information is out there about them. Look through the search engine results. Make note of any information you would like to make more private.

EXPERT QUOTE

As a victim of online defamation, I know firsthand what it is like to struggle with a tarnished virtual image. Fortunately, many people haven't gone through the malicious attacks I have been through—but today the slightest scarring on your virtual résumé can potentially eliminate you as a job candidate.

—Sue Scheff, author and family Internet safety advocate

- Privacy and Safety Housekeeping: Make a list of all the sites your kids visit. Go to each site and review privacy settings and change them if necessary to protect their privacy. Make a note to check privacy settings once a month, especially on Facebook, which typically assigns the weakest privacy settings by default when new features come out. Is your kid sharing too much information? For example, is he giving the year along with his birthday month and day, or has he mentioned what school he's attending?
- Keeping Up with Connections: How many connections does your kid have online? Does he or she have a Facebook profile with more than one hundred friends? Discuss with your children the consequences of having too broad a network and discuss ways to protect your child's information and reputation with those in their network who might not be trusted friends.
- Your Good Name: Buy a domain name with your kid's name in it. As your child reaches high school and college age, post appropriate information that would be helpful in establishing an online reputation. This could also include requesting references, and permission to post them, from teachers and part-time employers.
- Profiles in Search Results: Sites such as Twitter, LinkedIn, and Facebook all rank high in search engine results, so be sure to keep these profiles current and positive for children who are old enough to use these sites.
- You Are the Company You Keep: Even if you keep your children's profile private and positive, the friends they associate with online could reflect poorly on them. Talk with your children to make sure they understand this and help them choose carefully who they associate with online.

Get Creative

Creating a positive online persona can be a great project for you and your kids. This activity might include the following:

- When I grow up . . . : Have your children search online for people they admire. Talk about what that person did to get where he or she is in both professional and personal life.
- Motivational posts: Have your kids find a book or calendar with motivational quotes for the day. These can be great sources that they can use to start off their posts, texts, or e-mail messages.
- Subject matter expert: Does your kid have one school subject he or she really enjoys? Have your child post about that topic from time to time, even if it's just on his or her personal domain name page or on a blog you help him or her set up and monitor. Love learning Spanish? He could translate funny and favorite phrases.
- Occupational dreams: Does your kid want to be a sports anchor or a veterinarian when he or she grows up? Have him or her post information on a blog or social network about people who are in their dream job that he or she admires. Avoid uncalled-for criticisms or negative shout-outs.
- Creativity and innovation: Find outlets for your kids to show their creativity and innovation in a safe and nurturing way.

By having the conversation with your kids about Internet safety and making sure you are online where they are, you can help protect them and their reputation.

Interview with an Expert

We had the distinct honor to discuss the book concept, including the chapter dedicated to kids, with the distinguished Dr. Michele Borba, internationally recognized educator, author, and parenting expert on nurturing compassion and character and combatting bullying (with respect). She has written more than twenty-four books including her newest title, *Unselfie: The Surprising Role of Empathy in Preparing Children for Happiness and Success.*

Q: If you were talking to a parent today about their kids and the Internet and you wanted them to walk away with one message, what would that be?

A: You are always the best firewall to your child. Don't relinquish your power that you have as your child's parent when they venture online. I would tell parents to remember four things: Stay educated about the Internet. Know your computer. Know your child. And above all, stay in charge!

Q: On one of your *TODAY Show* segments, you talked about the digital age and its impact on kids. Can you describe the model parents who are handling their kids in the digital age well?

A: Model parents all have three qualities in common. Number one, the model parent is savvy. These parents realize that social networking online, via phone and computer, is here to stay and part of their child's life. Number two, the model parent is educated on the newest technology applications and devices. The third thing these role-model parents do is they parent the same online as they do offline.

You can reduce the risk factors for Internet problems if you are what I call a hands-on parent. Hands-on parents have regular conversations with their kids and discuss good behavior and what to avoid doing online and offline. They do not spy on their kids and have an open policy with their kids about how they will monitor them. They are in touch with the true maturity and trustworthiness of their kid and her friends and set clear boundaries and guidelines that apply to offline and online behavior. The model parent is also not afraid to say no.

Another technique you can use is to set the ground rules using the word *with* and treat the Internet and access as a privilege. When you set the rules, say things like "When you use your Facebook account to talk to your friends, I plan to be a friend with you and will read posts with you." You do not need to spy or wiretap them to keep up to date on their online activities; that could actually backfire on you. Make sure you openly let the kids know you will be where they are online. Let your kids know that you are committed to being online and savvy, too! Kids are saying they love to text and social network and they like to connect with parents this way. See technology as a challenge to undertake to stay in touch with your kids. Ask your kids to teach you.

Q: You have encouraged parents to focus on the issue of cyberbullying. What are some of the lessons you've learned from parents of both cyberbullies and victims?

A: The first thing is many of us incorrectly blame the Internet for all of this. At the core of the bullying problem is not the Internet; it is about how kids manage relationships. We know that the Internet is a big part of our children's lives. Our kids are plugged in a lot. Without the benefit of regular, face-to-face connections, kids may not develop the social relationship coping skills such as empathy, wanting to fit in, valuing differences, and respect for others. We also know from research that bullying peaks during middle school. You cannot wait until middle school to teach important social skills and how to manage relationships. That needs to start when they are a toddler and be well developed by the time they reach their tween years.

Q: What should parents watch for so they can be alert to any potential online problems?

A: There are some offline behavior clues that parents should watch for that may indicate there are online issues. The reality is that your kid may not tell you that something bad is going on. The trick is to watch your child's reactions in certain situations. Each situation is different, but there are some warning signs. Keep in mind that the signs may not indicate bullying or a predator relationship, but it should be checked out.

1. Have your child's Internet habits changed dramatically, such as a major decrease or increase in usage?
2. Has your child withdrawn from normal activities that they used to enjoy to spend more time on their smartphone or computer?
3. Is your child always trying to access the Internet when you're not there or from their bedroom?
4. Does your child get irritable or distracted when a phone call, voicemail, or instant message comes in?
5. Does your child receive strange phone calls, mail, or gifts from people you do not know? (A predator may send "gifts" to befriend a child.)
6. Does your kid switch screen names quickly or cover up the screen when you walk by the computer?
7. Has your child set up other accounts recently to receive e-mail or instant messaging?
8. Does your child appear nervous when you are using the computer?

Q: When we teach Internet safety classes to kids and parents, the message we try to get the kids to focus on is "digital is forever." Because a lot of kids have a hard time relating to "forever," what advice would you have for parents and teachers to help kids understand the concept that today's actions on the Internet could have long-term positive or negative results later in life?

A: We have spent a lot of time studying the baby years; we know all the social and physical milestones that children should meet from the time they are born until they reach the tween years. We really need to study the tween years. Tweens have a tough time pulling back and pausing before they act. Give them small steps and guide them along the way. Take simple steps like putting a little message on the computer screen that reminds them that there are no "take-backs" once you post something online. Remember, you taught them to look to the left and the right before crossing the street. It is the same online; teach them to take a deep breath, think, and then click. Give them a simple mantra: (1) no takebacks; (2) it is forever; (3) what they say today not only impacts them now, but it can also impact their friendships, ability to get a job, and ability to get into college later on.

Q: Focusing on the positive, what are some trends on the Internet that you see that are providing a positive and supportive experience for kids online?

A: I see some glorious trends. We are seeing this generation of kids becoming global. They can connect with children all over the world. They learn about other cultures, which leads to valuing differences and greater tolerance. They learn that kids have the same feelings as they do regardless of where they live. If we play our cards right, with instant access, we should have more time together. A dad on business travel can Web conference on tools like Skype so they can chat with their kids. A mom deployed in the military overseas can read *Goodnight Moon* to her child. Grandparents can stay connected more. [The Internet] is also expanding our kids' cognitive capabilities, with tools such as the new problem-solving games that expand our kids' minds.

A question was asked of high school seniors after taking the SATs if they would object to their parents friending them on Facebook, and most said they would not object, with many commenting they would welcome it. I think this is wonderful. The kids do want to be connected to us, and maybe, by connecting with them, we can build upon our relationship with them and interact with them more than was possible in the offline world.

CHAPTER 11

TURNING OFF THE LIGHTS
Choosing to Be Invisible Online

In the past, you may have opted for an unlisted phone number to maintain a sense of privacy. Today, an unlisted phone number won't protect you from people who are watching your every online action and observing your digital life. Every click, post, purchase, and search provides digital tracks that undermine your illusion of online anonymity.

In this age of Twittering, sharing your life and thoughts with your hundreds of online "friends" gives you many reasons to consider the alternative of becoming anonymous online.

Whether you post a picture of your latest hobby or click a "like" button on a listing for your favorite book, there is a downside to sharing your identity and making connections to others online: People are watching. Sometimes the watcher is your employer, your neighbor, or a prospective date trying to learn more about you. But keep in mind that commercial interests are following your every move and that bad guys are watching you, too.

Multiple personas can help you keep some aspects of your life private from certain people. You can build personas that reflect different parts of your personality and life favorably. You can also do this in a way that is authentic and genuine without acting like you have something to hide.

What Are the Benefits of Maintaining Your Anonymity Online?

The truth is people pursue online anonymity to protect themselves from unethical marketers, politicians, thieves, and other bad guys, and bad guys use anonymity to commit online crimes. This is a scenario that we will wrestle with for the foreseeable future.

In 2014, Apple and Google announced an upgrade of their smartphone operating systems that allowed a phone's user to encrypt the hardware so that not even the equipment manufacturer could bypass the encryption. The U.S. law enforcement community lodged immediate complaints, feeling that the police must have the ability to break into people's smartphones when they want to. Apple and Google refused to make a change to the hardware encryption protocols, and the battle continues.

This battle reached a breaking point when the U.S. Federal Bureau of Investigation petitioned, and a federal court granted the order, forcing Apple to prepare a new string of software code that would assist the FBI in breaking into the iPhone of a domestic terrorist. An appeals court reversed the order, and ultimately the FBI paid non-Apple hackers to provide the tools they claimed to need. This solution to an immediate problem does not mean that the entire issue is resolved, and the government is likely to push the phone manufacturer in the courts again.

Currently, law enforcement has access to more information about citizens than at any time in history. More of our lives than ever before are contained in digital files. For example, although some people still use cash for transactions, most payments are conducted with credit cards, debit cards, or through digital transfer methods, like your bank's bill pay application. Files are kept on each transaction. In cities large and small, cameras have been placed at nearly every intersection, allowing anyone with access to the cameras to follow your movements. In addition, your car and your smartphone are connected to the global positioning system, and several other systems that can pinpoint where you are at any given time. Some people even wear Fitbit wristbands or other wearable technology that tracks their location anywhere in the world.

And despite obvious concerns, U.S. law enforcement is adamant that access to all individuals' cell phones is necessary to fight crime and keep citizens safe. Anonymity is the enemy of law enforcement. Police forces want to be able to map the networks of criminals and potential terrorists, and they cannot do so if people in the chain can hide themselves enough to be completely anonymous.

Eric Schmidt, CEO of Google, wrestles with the distinction between privacy and anonymity: "Privacy is incredibly important. Privacy is not the same thing as anonymity. It's very important that Google and everyone else respects people's privacy. People have a right to privacy; it's natural; it's normal. It's the right way to do things."[1] While speaking on a panel covered by CNBC, he also mentioned that government might need to pierce the protections around anonymity. Schmidt said that "we need a [verified] name service for people. . . . Governments will demand it."

Cross-device tracking, discussed in chapter 1, is the enemy of anonymity. A website or telephone company may have identified which smartphone belongs to you, but if they do not know how to tie this identity to your laptop, your tablet, and your work computer, then you can preserve a level of anonymity. However, if current trends continue and marketers continue to find better ways of tracking your identity across your various Internet access devices, then your chance of anonymity disappears. Each new device will add to the complex file that marketers and telephone companies are keeping on you. When this happens, both your privacy and your option to remain anonymous are compromised.

There is a difference online between privacy and anonymity. When a company that does business with you online refers to privacy, they typically are referring to the laws and regulations that govern privacy. If you hear an executive talk about protecting customer privacy, the executive most likely is referring to protecting your Social Security number or credit card information.

The ability to transact and communicate anonymously goes much deeper than basic privacy. Anonymity means that you are avoiding the appearance of an online persona. To achieve real anonymity as a participant on the Internet, you would need to mask your Internet protocol (IP) address so that no one could track you, tracking the trail left as your browser moves from one site to another.

Currently, anonymity is defeated because a site can identify the IP address from which an Internet user enters that site. If this address points to a home or a company, then the Internet user can quickly be identified. Even people using public networks in places like the local library or Starbucks can often be tracked because the public network may require an identification card to use its system, or the business offering the network is likely to have security tapes from cameras in the shop.

Law enforcement doesn't like the idea of anonymity because it means that criminals can cloak their Internet transactions and make them harder to identify. We feel that you have a right to anonymity online. We also feel that you have a right to manage your online persona as you wish, not according to how a marketing company or a data aggregator chooses to analyze you. We wrote this chapter not to provide complete instructions on staying anonymous. Although there are tools discussed here that can help you approach that goal, we assume that most of our readers simply want to become safer and more private online, so we propose methods to allow you to hide your identity better.

What's important is that you know how to manage the level of privacy and anonymity that is right for you to protect yourself online.

Avoid Identity Theft

Identity theft involves gathering and using information about your personal life to pose as you, usually with a goal of stealing your money. Identity theft can take many forms. Most often a person wants to use your identity to hide behind your good name, to say mean things, or commit criminal acts.

A 2015 Javelin Strategy and Research report found that 12.7 million people were victims of identity fraud in the United States during the previous year and $16 billion was stolen from consumers. The U.S. Federal Trade Commission has reported that identity theft was the top consumer complaint in 2015 and for the last fifteen years. Remaining anonymous online, or as anonymous as possible, can help you to avoid becoming a victim of identity theft.[2]

One major risk of becoming a victim of identity theft involves data breaches at companies. Sony PlayStation Network was hacked and some experts believe that personal information for more than 77 million Sony PlayStation Network customers was taken. Though the thieves did not get credit card data, they may have stolen the building blocks they needed to steal identities such as names and e-mail addresses. More recent famous hacking cases such as those at Target, Home Depot, and Neiman Marcus caused millions of consumer records to be exposed to potential identity thieves. It is likely that such theft will continue to get worse; for example, there has been a recent uptick in hacking attacks seeking patient information in the healthcare industry.

Security experts have shown that they can assemble information to re-create your identity at lightning speed based on what you, the government, and others post about you online. These experts can take educated guesses at your password as they look at your Facebook profile and see your pet's name, favorite sports team, and the year you graduated from high school.

Look at your profiles, posts, and pictures carefully. If you have posted anything online that could be used to impersonate you, remove it. For example, you can keep your birth month and day on Facebook, but don't display the year of your birth or your age. Many people post their phone number, e-mail address, and hometown. The truth is that much of this is public information, but you should still make the bad guys work to get it. There is no reason to provide them with a one-stop shop for all the information they need to impersonate you.

Don't Become a Social Engineering Target

Social engineering is a trick of the trade these days for Internet fraudsters and criminals. A person will troll the Internet for your personal information and then use that information to contact you to trick you into thinking he or she is someone who can be trusted. A social engineer may call you on the phone posing as a florist and tell you that you are receiving flowers from your husband but that they got his credit card number wrong and cannot process the order, so you provide your card number. They may e-mail you posing as your bank, explaining that there is a problem processing an automated payment and asking you to click on a link to verify your account information. By sharing as little information as possible, you can keep yourself a little more anonymous than others and avoid some social engineering attacks. A competitor to Facebook, Google+, allows you to set up circles of people, which might be a great reminder. When you decide to post something, whether a news article or the latest photo from your exciting cruise, Google+ has an interface that makes it easy to see that you are going to make a public post, or it allows you to choose circles of people that you defined.

This is a great way to make sure you do not broadcast where you are, or where you are not, to the social networking public.

When a major marketing company, Epsilon, was hacked and names and e-mail addresses of customers were taken, we waited to see what types of social engineering attempts would ensue. Epsilon sent roughly 40 billion e-mails last year on behalf of roughly 2,500 companies. That is a lot of e-mail addresses, names, and other demographic information. Companies such as the Ritz-Carlton, Best Buy, and Capital One were customers of Epsilon at the time of the breach, and it appears that their customer accounts may have been accessed. One of the authors of this book tracked several scam e-mails from the companies involved that looked incredibly convincing.[3] The e-mails used online nicknames and the special e-mail address used for reward programs, which provided a clue that these e-mails were not legitimate. When e-mails asked the recipient to click on a link and update personal information, it was a sign of a social engineering attack.

Social engineers also gather information about you that they can use to cyberbully you online. Cyberbullies target an individual and threaten or taunt him or her using text messaging, e-mails, and posts on social networking sites. One benefit of maintaining anonymity, especially for those younger than twenty-one years of age, is that you avoid providing information to bullies. Some cyberbullies take over the account of the intended victim or create a phony account and post embarrassing photos or other content. In a deeply saddening case of social engineering and cyberbullying, Megan Meier committed suicide. Her parents believed that their daughter's suicide was directly related to having being taunted online. During an investigation into her suicide, it was found that a bogus account was created on Myspace under the name of Josh Evans. Megan was sent e-mails and messages by several people using the fake account. During a court trial, witnesses said that the purpose of the bogus account was to trick Megan, or socially engineer her, into giving up information that could be used later to humiliate her.[4]

Controlling Your Online Presence

When you hear about the site Facebook, you think pictures, videos, and fun posts from family and friends. Facebook counts more than 800 million active users in their community. LinkedIn is a site for professional networking. As of early 2015, LinkedIn claimed more than 332 million users across more than two hundred countries around the world and in twenty languages; the site receives 187 million unique visits every month. LinkedIn has seen explosive growth, as much as 400 percent over in some years, in their user base accessing the site via their mobile phones.

Many professionals say it is not that easy anymore to control your online presence and keep your professional and personal lives separate online. Most people, when pushed to admit it, have a blended and blurred mashup of their professional and personal lives online. However, you can define the line between the two.

Separate Private and Professional Lives

Because your life is online for anybody to see, you may want to establish a division between your professional and personal online personas. If you get a friend request on Facebook from someone you'd rather keep on the professional side of your life, there's nothing wrong with not accepting that request and sending the person a LinkedIn invite. You may also consider creating a Twitter account for your personal life and a separate one for professional tweets, as long as you know that anyone can follow both of your profiles and tweet accordingly. If you do friend people you work with on a social networking site, you can use your privacy settings to control who can see your posts and pictures by going to the site's sharing settings (on Facebook, for example, use the "Sharing on Facebook" section).

As a first step, refrain from using your work accounts for personal activities; if you don't, you may expose your personal activities to your boss and coworkers. For example, the online dating website, Plenty of Fish, was hacked, exposing usernames and addresses. Typically, vendors have to notify the domain names when they are hacked and provide a list of names within that domain. Would you want your company to see Bob.Smith@ABCCompany.com was on the list? Another recent example is the hacked gaming platform for the Sony PlayStation. If you like to play online games on the weekend on your Sony PlayStation but you used your work e-mail address to sign up, your employer might know. If the PlayStation is breached, various domain names for e-mail accounts would be notified so the domain names could notify their users as well as to keep them alert to potential e-mail scams.

You can also create e-mail addresses and nicknames that you only use when you sign up for rewards programs. These sites could be breached, but they have different notification standards from many retail or banking sites and may not tell you about the breach right away. If you create these special-use e-mails, you will be able to more easily identify e-mail scams that come to these accounts and the sites that may have generated them.

Purge and Protect

Review all e-mail mailing lists you are currently on. Unsubscribe to the ones you don't read regularly to save you time managing your inbox and to shrink your Internet footprint.

Review your Facebook and other social networking sites. Either purge the number of connections or use privacy settings to protect what people can see about you. You should also think about closing e-commerce accounts that you no longer use, especially if they have your credit card or bank information on file. If you are not using the accounts, it makes it easier for someone to take over the account without your knowledge or for them to send you scam e-mails trying to trick you into clicking on links or opening attachments. Sign onto the site you want to cancel and type in search terms such as "deactivate," "close account," or "delete account." For example, if you want to close your account on Amazon, you must first sign in and select the "Contact Us" button. Once there, you pick "e-mail" and then select "Close My Account." The deactivation is not instantaneous, and some sites mention that it could take up to one to two weeks before the account is closed.

Stay Safe

You might have a reason to remain somewhat anonymous for your personal safety. You may be signing up with a dating site and you don't want bad guys stalking you. You may have recently left an abusive relationship and you want to be able to surf the Net and use your smartphone and not have your former partner track you down.

If your safety is a primary concern, consider being offline as much as possible. If your work requires you to be online, look for opportunities to maintain

---

QUICK TIPS FOR BEGINNERS ON BECOMING MORE ANONYMOUS

If you only have twenty-five minutes to start establishing your anonymity online, here are four steps you should take going forward:
1. Avoid using your real full name online for account IDs and e-mail IDs.
2. Ask yourself why a site is asking you to reveal personal information for your online accounts and opt out of providing this information whenever possible.
3. Set up a separate account and nickname for use on sites where people with like interests gather (called *affinity groups*).
4. Talk to your kids or other family members about what they can safely post or discuss online.

anonymity and do not make careless mistakes that can jeopardize your anonymity or your safety. In several cases, an error in judgment allowed the bad guys in, either because a person was in a hurry and had to do banking on a public WiFi or was distracted when she clicked on an e-mail scam that installed malicious software on her computer. Plan your time on the Internet wisely. If you are in a rush, refrain from starting any sensitive transactions until you have more time to conduct them thoughtfully and carefully.

Choose Names That Don't Reveal You

Just as Samuel Langhorne Clemens used the pen name Mark Twain, you can use names that don't reveal exactly who you are. Using names other than your own doesn't mean you are completely anonymous, but it does make it a little harder for people to guess your real identity.

There are two situations in particular where Internet users may want to consider not using a real name: dating sites and sites used by children younger than eighteen. Especially for children younger than age eighteen, it's best to avoid using a real name, a nickname used in real life, anything that identifies the minor as male or female, and anything that identifies age or school.

Here are a few naming tips:

- Use a name that is more generic, such as "SmithFamily" instead of "SallySmith."
- Consider using your initials along with an activity you like to do such as SSTennis.
- If you are having a tough time coming up with a name, try a tool such as Username Generator at www.usernamegenerator.net to help you pick a name that you can remember.
- Look up your name or hobbies in another language and use the foreign words for a user name.
- Stay away from unprofessional sounding names or create a separate account with a more professional name just in case you need to use that account in a work setting, such as applying for a job. Using Sexymama@yahoo.com or Luvs2Party@hotmail.com might not portray the image you would like on the job application.

Read the Privacy Statement

According to a Nielsen survey released in April 2011, more than 50 percent of those surveyed, both males and females, have privacy concerns when it comes to location sharing on their mobile devices.[5] Every time apps share your location, you become less anonymous. When location sharing is active, your whereabouts, habits, and the places you frequent are all collected and assembled into patterns.

Each of your app providers or companies that activate and use your smartphone's location information also publishes privacy policies to tell you what data they take and what they do with the data. Yet almost no one reads these privacy policies. We could decide whether to use a mobile app with full information about how our location information will be collected and used, but we tend to ignore that information. Reading a privacy policy can tell you about the operations of an app, and reading several of them allows you to compare which companies treat your privacy more seriously, and are less eager to share your data with others.

Opera Software asked consumers in Russia, Japan, and the United States about privacy on the Internet. Each group selected Internet fraud as the result of a breach in privacy as being at the top of their list of online worries. In fact, Internet privacy actually beat out worries about terrorism and going bankrupt in the survey.[6] Still, how many people actually read those privacy policies that pop up when you sign up for a new account? For the top one thousand websites, more than one third will offer you a link to the networking advertising initiative to opt out of tracking. More than 10 percent explicitly say they will share your information with third parties.

Yes, the privacy policy looks legal, long, and tedious to read. According to an info graphic posted on Mashable.com, the average privacy policy is 2,462 words long.[7] A policy could take you roughly ten to twenty minutes to read, but the time will be well spent.

Cloaking Tools: Anonymizers and Remailers

Criminals are good at being anonymous online. They use various tools and tricks to hide their misdeeds and make it harder for law enforcement to catch them. But, some of those tools can be used by good and decent people to protect their identities online. You can become more anonymous online and better manage your online personas, but you will probably never erase yourself from the Internet. In our Internet safety class for K–8 kids, we make the kids repeat the mantra, "There is no such thing as anonymous on the Internet." This is essentially a true statement. Given time, money, and the right technical resources, you can be traced by savvy cybersecurity sleuths, criminals, spies, or law enforcement. For that reason, you may want to consider using an anonymizer or remailer to protect your privacy.

Anonymizers Defined

An anonymizer is a tool that can help make your Internet hops and searches harder to trace back to you. Activists around the world who live in countries that do not support free speech often use anonymizers to help protect their identities because they e-mail, blog, and tweet the truth about their country's actions to the world.

Usually an anonymizer involves a third-party website that acts between you and the site you visit. For example, if you decide you want to throw behavioral tracking software off your digital trail, you could use an anonymizer before you go shopping. When you want to visit Amazon.com, the browser connects to the third-party website using the anonymizer first. Once the anonymizer sees your request, the third-party website takes you over to Amazon.com. Your first hop is not easy for Amazon.com to "see"; they only recognize that the request came from the anonymizer.

Good people with good intentions may want to use anonymizers. For example, a prominent executive may use this technology while researching personal health issues. A doctor may want to look up information about sexual addictions on his patient's behalf. A professor conducting research on a controversial topic may want to do so anonymously. You may want to consider using anonymizers to keep behavioral tracking companies from following your every move online.

Law enforcement may use anonymizers, for example, to conduct online surveillance. They can surf sites or even inquire about services online without leaving digital clues that could put an undercover operation in jeopardy. Anonymizers also allow law enforcement to offer anonymous tip lines where citizens can report tips without fear of being traced.

Remailers Misrepresent You

Think of the return address you put on an envelope that you send through the postal service. A remailer is something like using a false return address so the recipient doesn't know who sent the letter. Remailers are servers that can receive e-mails and then send them on to the final destination without revealing the original source. The remailer may actually change the information in what is known as the e-mail header address to give a fake source address. If you want to avoid being identified by a company you e-mail to avoid receiving annoying marketing notices, or you want to avoid being associated with a particular group, remailers can come in handy because they move, or remail, your information and prevent the receiver from tracking you as the source or tracking your location information.

Anonymous Only Goes So Far

Sophisticated security experts can track down anonymizers. Anonymizers and remailers offer handy tools to help protect you, but they are not foolproof from bugs or from other tools that could eventually trace traffic back to you. More than likely, if you are using anonymizers or remailers, they'll protect your identity. However, keep in mind that you do leave digital footprints behind even when using these techniques.

For example, if you attach files to an e-mail sent through a remailer or anonymizer, the file you created most likely embedded the name, product serial number, and the computer ID of the computer the message was created on. In addition, the intermediate machine may leave some clues, depending on how sophisticated the service is. Also, if you sent or received the message using a pretty good privacy (PGP) digital signature (which can be used to sign and encrypt your Internet e-mails, text messages, and even documents), the PGP digital signatures can offer clues that will reveal who you are.

Stop Others from Stripping You of Anonymity

Technology is great, but the best way to stop others from stripping you of your anonymity is to modify your personal behavior. Each time you click, sign up for a program, or join an online community, you are giving up pieces of your anonymity. You can still enjoy all the benefits the Internet has to offer, but you will have to manage your online activities carefully.

Mobile Device Recycling

Each year we buy new devices that leave us with a pile of digital junk to get rid of. Many people like to use trade-in services or even resell their devices on auction sites such as eBay or Craigslist. You might want to think twice before you hand your mobile device to a stranger. When you sell, give away, or throw away a mobile device, you can hand over the keys to your digital life.

A recent study found that roughly 54 percent of used phones sold online still contained sensitive data. Those data included credit and debit card details, PINs, passwords, address books, and more. In the wrong hands, a mobile device can provide access to your browsing history and your contacts, and may even allow someone to open up your e-mail and social networking apps to snoop on your digital life.

The best way to avoid this scenario (other than dropping the device into acid and then running it over with a tractor) is to call the manufacturer of the device and ask for advice on how to permanently wipe the data it contains.

Many people think that deleting the data is sufficient, but it isn't. Steps you need to take to erase data from mobile devices typically include the following:

1. Log out of every application and delete each app.
2. Use the permanent wipe function as directed by your manufacturer.
3. If possible, remove the SIM chip.

If you are unsure how to go about this, ask a technical professional for assistance. We also recommend turning the device into a reputable recycle program. Alternatively, you can keep your "digital junk" in a drawer in your house and use the devices as backups for your newest digital devices or for extra storage.

Settings Can Protect You from Prying Eyes

Computer and software manufacturers have put several key safety settings in place to help you stay safe out there on the Internet. You should update your computer operating system regularly, as such updates often include security fixes. Similarly, use the most current version of browsers and update your antivirus and antimalware software frequently.

There are several technology tools you can use to help you maintain some anonymity when you're online.

- Cookie alerts: Set your browser to alert you every time a site tries to install a cookie (a small program that stores information about your browsing history) so you can choose whether or not the cookie is installed.
- Strong passwords: Using strong passwords can help you protect your anonymity online by keeping the bad guys out of your social networking, e-commerce, and e-mail accounts.
- Deleting tracking cookies: Tracking cookies can be used to follow your online activities, possibly revealing the passwords and account information you enter on your keyboard to others. Set up your antispyware program to delete tracking cookies.
- Deleting browsing history: Most popular browsers offer an option to delete your browsing history so others can't easily see where you've been online.
- Firewalls: Use a firewall in your operating system or from a third-party software program to protect your home Internet access.
- E-mails: When you sign up for e-mail alerts from a company, choose "plain text e-mail" instead of HTML. You don't get the fancy graphics, but you avoid the cookies typically sent in HTML messages.
- E-mail account: Make sure you use your e-mail service provider's SSL-encrypted option. Many of these options are newer, and you might have to opt in to get this service. By using the SSL-encrypted service you help keep snoopers and prying eyes from stripping you of your anonymity.

Revamping Your Social Networking Habits

Social networking was designed to reveal and connect you to others. If you only want to share information with a few people, social networking sites are not the place to share. To take charge of your online identity, it's important that you review your social networking habits.

Look at what information you share online about yourself and your loved ones.

- Review your privacy and security settings regularly. Keep in mind that even when you lock down your settings, new features and functions might bypass those settings.
- Read privacy statements so that you know exactly how much of your anonymity you still control and what you are giving up when you join sites such as Facebook, Twitter, LinkedIn, or Myspace.
- Remove postings that reveal information such as your full legal name, the year you graduated from high school, your full birth date, and full home address.

Facebook's Happening Now feature shows you what people like at the moment, as well as photos and information that have been posted online. With this service, even if you post a photo and your settings are set to be viewable by friends only, the photo might still get picked up in the Happening Now stream.[8]

Researchers have found a privacy snafu on Twitter in a feature called direct message (DM). Using this feature you can send a direct message between you and another person. This feature is almost like cell phone texting but within the platform of Twitter.

Unfortunately, most Twitter users who were early users of the system were unaware that a programming bug allowed third-party apps to read your direct messages.[9] Reminiscent of the party line in the early days of the telephone, this glitch was an open invitation to eavesdroppers.

Losing Anonymity by Accident

Here's an interesting case where somebody gave up almost all anonymity while using a social networking site. A young girl who was turning sixteen was planning a birthday party. She set up an event on Facebook, including her date of birth and home address. She sent a notice of the event to her network of friends. The problem was the girl missed one important feature on Facebook. When you send an event to "Anyone," it doesn't just go to anyone in your personal network, it goes to anyone on Facebook. In other words, anyone, means everyone. Roughly fifteen thousand people sent an RSVP saying they would attend the party. Even though she later canceled the party, approximately 1,500 people still showed up.[10] Now the whole world knows this young lady from Germany, her name, and the fact that she just turned sixteen. Her family

FACEBOOK FACIAL RECOGNITION

Facebook uses facial recognition technology to allow your friends to tag you in a photo. The ultimate goal is to automatically tag you based on photo recognition technology without relying on your friends to tag you. You can turn this feature off by going into Facebook and changing your account's privacy settings. Under "Things others share," go to "Suggest photos of me to friends" and edit the settings to "Disabled" if you do not want to be tagged by others.

literally had to run away from their house on her birthday to make sure that they were safe from party crashers.

Account Safety Online

Use accounts that offer https options for e-mail and other accounts whenever possible. "Https" stands for Hypertext Transfer Protocol Secure. You may see https in the name of the site URL when making secure payments online. Using https for your e-mail and social networking profiles will protect your security and, ultimately, your anonymity from creepy cybercrooks and snoops.

If you would like to turn on the https standard, you can search for instructions within the help feature of the site on which you have an account or any site you visit.

If you have a new smartphone or digital camera, chances are it has a fun feature built in called *geocoding*. Geocoding automatically saves the latitude and longitude of your location when you snap a photo. Several social networking sites also allow you to post your location along with your content. You should turn off your geocoding or location-based information when you send posts so you do not broadcast where you are. If you want to turn off this feature on your phone or camera, look for options in the device's settings. When in doubt, contact the manufacturer and ask for help.

Alternate Identities: Choosing to Be Somebody Else

"On the Internet, nobody knows you're a dog" was the caption of a cartoon by Peter Steiner published in the *New York Times* on July 5, 1993.[11] That sentiment

WHY WHISTLEBLOWERS AND ACTIVISTS ARE USING TOR

When maintaining anonymity is critical, a tool called Tor can help protect your privacy while surfing online. Tor is designed to enable people to communicate safely using the Internet. Tor routes traffic through a volunteer network of servers to help maintain your anonymity. Tor came from the U.S. Naval Research Laboratory and has evolved over time. Tor is now a nonprofit organization based in the United States, and the tool will work on Windows-, Mac-, or Linux-based devices. Tor is popular with activists in countries that do not support free speech. Go to https://www.torproject.org/projects/projects to learn more about Tor.

is still relevant today. You can create an alternative identity or ego online, and many people do. Perhaps you need to protect your real identity from others, or you may want to present different images of yourself online such as your professional image, your personal image, and your image as a member of a particular organization. Whatever your motivation, there are some things you should know before creating alternative online personas.

Risks of an Alter Ego

An alter ego can be difficult to maintain. You have to remember what you are sharing, what e-mail address you used, how you answered security questions, what image you projected in the content you posted, and more.

There is also the danger that someone will link your alter ego back to you, which could lead people to ask you why you created an alter ego in the first place. Some may not be able to trust you fully again. In addition to those risks, you can also find yourself in violation of user agreements because you provided fictional user information. Violation of user agreements can get you kicked off sites, or worse.

For example, Facebook does not allow fake user names, according to user policy. Facebook and others do this to prevent spammers and from using bogus accounts to trick customers into clicking on bad ¹ away personal information. This policy has led to issues with changed their names, and some people who identify as a the one they were born into. Some people have developed a pseudonym to conduct all of their business, but use a different name for their Facebook accounts.

It is estimated that, even with such a policy in place, there are roughly more than 25 percent of accounts on Facebook that are deemed bogus or fake. This matches up with a survey called the "Cyber Norton Report: The Human Impact,"[12] which found that roughly 17 percent of respondents lied about age, financial status, or marital status when they were online.[13]

How Sock Puppets Help the U.S. Military

Alternate identities can be a useful way to protect your anonymity while online. You may have a connection to the place where you work, but outside of work you are exploring a career change or want to establish your own professional brand. Perhaps you have different sets of friends or volunteer organizations that you want to see different sides of you.

The U.S. military has intelligence operations that use fake online identities to protect the real identity of a person collecting information. The *Huffington Post* ran a story about how the military is working with a software provider to help their personnel manage multiple identities, all created for online military missions.[14] Using these fake identities, operatives are able to befriend the enemy and collect information to help avoid conflicts and human casualties. This device for using false personas is called *sock puppets*. Sock puppets are online identities used to promote ideas or to gather intelligence without revealing the true identity of the person behind the fake identity.

An Alter Ego Makes a Divorce Worse

Alter egos can help you build a little anonymity across the facets of your life. But sometimes, creating alter egos is a bad idea, especially if you are creating them with the intent of snooping on somebody else. Angela Voelkert was a twenty-nine-year-old going through a divorce. She decided that she wanted to anonymously snoop on her soon-to-be ex-husband, so she decided to create a fake persona.

Angela used another name, Jessica Studebaker, and said that she was seventeen years old. As part of her snooping, she contacted her husband on Facebook. Using the fake profile, she befriended him. As the friendship progressed, David asked "Jessica" to go away with him. He also admitted that he had installed a GPS tracking device on his wife's vehicle. He told Jessica he wanted his wife dead, asking Jessica if anybody at her school would want to commit murder for $10,000. Law enforcement has since become involved and determined that David knew it was his wife faking it all along, and he was baiting her.[15] This bizarre story of a problem marriage bears an important reminder about the perils hiding behind a fake persona.

Alter Ego Can Make for Professional Embarrassment

Exposure of your alter ego may reflect poorly on you. People might wonder why you hid behind the alter ego to begin with. Consider John Mackey, the Whole Foods CEO who was posting on Yahoo! message boards under a fake name. Using the online handle of "Rahodeb" (his wife's name spelled backward), he was posting positive press about the company as if he were unconnected to company management. In one *Wall Street Journal* report[16] it was noted that under this alter ego, Mackey would say wonderful things about Whole Foods while trashing a competitor that his company later bought. There were allegations that he was trying to lower the price of the competitor's stock. While the merger was in its early stages, his alter ego came to light and the Federal Trade Commission wanted an explanation. You need to think twice before you post under what you deem is a cloak of anonymity.

There is also a potential downside to anonymity. One marker that law enforcement and anti-terror investigators look for when they collect information is people who have tried to cloak their activities. If you are using tools to make yourself anonymous online, you may be attracting the attention of law enforcement. Police know that people planning criminal activity try to cloak their actions if possible, and so the use of these anonymizing tools can catch the eye of police scanners. It is ironic that the very actions you take to hide from police are the same actions that could bring you to their attention.

Conclusion

When we first decided to write this book, we both exclaimed that we wanted to write a book for our moms, grandmoms, friends, and kids. It was our passion to help others have fun on the Internet and to teach them the tips and tools needed to build, protect, and enhance their Internet identity.

Microsoft founder Bill Gates is quoted as saying, "The Internet is becoming the town square for the global village of tomorrow."[17] Your Internet identity is your storefront on this town square. You may see news reports about the dangers of being online that make you want to hide and avoid it. Don't do that. If you fold up shop and refuse to participate, you will leave a void where your Internet presence should have been. If you let the storefront become shabby and overgrown or let people trespass on the storefront, people will notice. If you are advertising the wrong things, it may come back to haunt you.

Being on the Internet does not have to be daunting and scary. There are so many ways to constructively participate in our electronic global village, and to do so from the comfort of your own living room, that it would be a shame to miss these opportunities. By taking to heart the lessons in this book, we hope you can build, grow, and maintain the most appropriate Internet image for you. As the Internet weaves ever deeper into our society, you are now well equipped to protect and defend your online persona and to take full advantage of the Internet's benefits. Enjoy your time online, and keep polishing your Internet identity.

NOTES

Chapter 1: How Were You Exposed?

1. April Witt, "Blog Interrupted," *Washington Post*, August 15, 2004.

2. Roxanne Roberts and Amy Argetsinger, "Jessica Cutler: From 'Washingtonienne' Scandal to New Mom," *Washington Post*, October 23, 2009; Matt Apuzzo, "'Washingtonienne' Blogger Filing for Bankruptcy," Associated Press, June 1, 2007.

3. Michael Thomsen, "PewDiePie Doesn't Make Anywhere Close to What He Should Be Making," Forbes.com, July 11, 2015, http://www.forbes.com/sites/michaelthomsen /2015/07/11/pewdiepie-doesnt-make-anywhere-close-to-what-he-should-be-making.

4. David Wright, Chris Murphey, and Lauren Effron, "Meet the Vine Stars Who Turn 6 Seconds of Fame into Big Bucks," ABC News, September 15, 2014, http://abcnews .go.com/Business/meet-vine-stars-turn-seconds-fame-big-bucks/story?id=25522189.

5. Chris Jay Hoofnagle, Jennifer King, Su Li, and Joseph Turow, "How Different Are Young Adults and Older Adults When It Comes to Information Privacy Attitudes and Policy?" Social Science Research Network, April 14, 2010, http://papers.ssrn.com/sol3/ papers.cfm?abstract_id=1589864.

6. Dave Eggers, *The Circle* (New York: Alfred A. Knopf and McSweeney's, 2013), 441.

7. Jenna Wortham, "More Employers Use Social Networks to Check Out Applicants," *New York Times*, August 20, 2009.

8. "Big Surge in Social Networking Evidence Says Survey of Nation's Top Divorce Lawyers," American Academy of Matrimonial Lawyers, February 10, 2010, http:// www.aaml.org/about-the-academy/press/press-releases/e-discovery/big-surge-social -networking-evidence-says-survey-.

9. Amy-Mae Turner, "8 Things You Really Should Know about Facebook Photos," *The Daily Dot*, February 2, 2016, http://www.dailydot.com/technology/how-to-upload -facebook-photos-help.

Chapter 2: Peekers and Gawkers

1. "The Monitor's View: How Europe, US Can Solve Internet Privacy," Monitor's Editorial Board, October 6, 2015. http://www.csmonitor.com/Commentary/the-monitors -view/2015/1006/How-Europe-US-can-solve-Internet-privacy.

2. Sam Thielman, "Privacy Groups Hail 'Freedom from Surveillance' in European Court's Facebook Ruling," *The Guardian*, October 6, 2015. http://www.theguardian.com/business/2015/oct/06/europe-court-right-to-privacy-max-schrems-us-tech-companies.

3. Sunil Patil, Bhanu Patruni, Hui Lu, Fay Dunkerley, James Fox, Dimitris Potoglou, and Neil Robinson, "Online Privacy vs Surveillance: Europeans' Preferences on Internet Surveillance and Security Measures," Rand Corporation, 2015. http://www.rand.org/content/dam/rand/pubs/research_briefs/RB9800/RB9843z2/RAND_RB9843z2.pdf.

4. Mary Madden and Lee Rainie, "Americans' Attitudes about Privacy, Security, and Surveillance," Pew Research Center, May 20, 2015. http://www.pewinternet.org/2015/05/20/americans-attitudes-about-privacy-security-and-surveillance.

5. David Anderson, a specialist on EU law, in a report for the British government, "A Question of Trust, June 2015."

6. Mark Penn, "Views from around the World, 2nd Annual Poll on How Personal Technology Is Changing Our Lives," Report for Davos, Switzerland, January 2015. http://mscorp.blob.core.windows.net/mscorpmedia/2015/01/2015DavosPollFINAL.pdf.

7. Katie Halper, "A Brief History of People Getting Fired for Social Media Stupidity," *Rolling Stone*, July 13, 2015. http://www.rollingstone.com/culture/lists/a-brief-history-of-people-getting-fired-for-social-media-stupidity-20150713.

8. https://www.nlrb.gov/rights-we-protect/protected-concerted-activity.

9. Erik Palm, "Facebooking While Out Sick Gets Employee Fired," CNET News, April 27, 2009.

10. 2007 Electronic Monitoring & Surveillance Survey from American Management Association (AMA) and the ePolicy Institute.

11. "2014 Social Recruiting Survey by Jobvite.com." https://www.jobvite.com/wp-content/uploads/2014/10/Jobvite_SocialRecruiting_Survey2014.pdf.

12. Jennifer Waters, "Facebook Is Fun for Recruiters, Too," *Wall Street Journal*, July 24, 2011.

13. U.S. survey by CareerBuilder.com, http://www.careerbuilder.com/share/aboutus/pressreleasesdetail.aspx?sd=5%2F14%2F2015&id=pr893&ed=12%2F31%2F2015; UK survey by Protecting.Co.Uk, http://protecting.co.uk.

14. http://www.fastcompany.com/3046133/the-new-rules-of-work/the-future-of-workplace-surveillance.

15. Melanie Hicken, "Big Data Is Secretly Scoring You," CNN Money, April 2, 2014. http://money.cnn.com/2014/04/02/pf/consumer-scores/index.html?iid=EL.

16. Alex Rosenblat, Tamara Kneese, and Danah Boyd, "Networked Employment Discrimination," Data & Society Working Paper, October 8, 2014. Prepared for Future of Work Project supported by Open Society Foundations. http://www.datasociety.net/pubs/fow/EmploymentDiscrimination.pdf.

17. Julianne Pepitone, "6 Painful Social Media Screwups," CNN Money, April 7, 2011.

18. Ellie Zolfagharifard, "Are YOU Being Watched? Walmart's Use of Facial Recognition Software to Spot Thieves Raises Privacy Concerns," *Daily Mail*, November 10,

2015. http://www.dailymail.co.uk/sciencetech/article-3311264/Are-watched-Walmart
-s-use-facial-recognition-software-spot-thieves-raises-privacy-concerns.html.

19. "Ohio Woman Says She Discovered Husband's 2nd Wedding on Facebook,"
Associated Press, August 5, 2010.

20. IBM Security Services Report, Cybersecurity Intelligence Index, 2014.

21. Paul Ferrillo, Weil, Gotshal & Manges LLP, "Changing the Cyber Security Playing Field in 2015," Verizon and US Secret Service Data Breach 2014 Report, January
20, 2015.

22. Symantec, Internet Security Report, 2014; Kim Zetter, "Hacker Lexicon: A Guide
to Ransomware, the Scary Hack That's on the Rise," September 17, 2015. http://www
.wired.com/2015/09/hacker-lexicon-guide-ransomware-scary-hack-thats-rise.

23. Zetter, "Hacker Lexicon."

24. Uptin Saiidi, "How That Angry Tweet at a Company Could Get You Hacked,"
NBC News Blog, October 24, 2015. http://www.nbcnews.com/tech/tech-news/how
-angry-tweet-company-could-get-you-hacked-n450686.

25. David Goldman, "Apple Bans Hundreds of iPhone Apps That Secretly Gathered
Personal Info," CNN Money, October 19, 2015. http://money.cnn.com/2015/10/19/
technology/apple-app-store/index.html.

26. Nathan Olivarez-Giles, "Android Malware Removed from Google Play Store
after Millions of Downloads," *Wall Street Journal* Blog, February 14, 2015. http://blogs
.wsj.com/personal-technology/2015/02/04/android-malware-removed-from-google
-play-store-after-millions-of-downloads.

27. Sasha Brown-Worsham, "Sexting: What You Need to Know," March 2015,
NJFamily.com blog. http://www.njfamily.com/Raising-Teens/Sexting-What-you-Need
-to-Know.

28. NCMEC, http://www.missingkids.com/KeyFacts.

29. Caroline Davies, "Revenge Porn Cases Increase Considerably, Police Figures
Reveal," *Guardian*, July 16, 2015. http://www.theguardian.com/technology/2015/jul/15/
revenge-porn-cases-increase-police-figures-reveal.

30. Loulla-Mae Elefheriou-Smith, "'Revenge Porn' Criminalised: What Is It and
What Are the Consequences?" *Independent UK*, September 23, 2015. http://www.ind
ependent.co.uk/news/uk/crime/revenge-porn-criminalised-what-is-it-and-what-are
-the-consequences-10042291.html.

31. Victor Luckerson, "Google Will Remove Revenge Porn from Search Results,"
Time, June 19, 2015. http://time.com/3928830/google-revenge-porn-remove.

32. https://newsroom.fb.com/news/2015/11/global-government-requests-report-4.

33. Glyn Moody, "Microsoft Wants US Government to Obey EU Privacy Laws,"
Ars Technica, October 21, 2015. http://arstechnica.com/tech-policy/2015/10/microsoft
-wants-us-government-to-obey-eu-privacy-laws.

34. Kim Zetter, "Caught Spying on Student, FBI Demands GPS Tracker Back,"
Wired, October 7, 2010.

35. Athima Chansanchai, "Tweet Costs Chinese Woman a Year in Prison," Techno-
log, MSNBC.com, November 18, 2010.

36. Tom Parfitt and Chris McGreal, "'Spy Swap' Under Way as 10 Plead Guilty in US Court," *Guardian*, July 8, 2010.

37. Taylor Buley, "Friending a Spy on Facebook," Forbes.com, June 29, 2010.

38. Brad Heath, "Police Secretly Track Cellphones to Solve Routine Crimes," *USA Today*, August 24, 2015. http://www.usatoday.com/story/news/2015/08/23/baltimore -police-stingray-cell-surveillance/31994181.

39. The source doesn't want to be named.

40. State of Privacy Report, Symantec, 2015.

41. Tess Danielson, "DuckDuckGo CEO Calls Out Google and Says It's 'a Myth You Need to Track People to Make Money,'" *Business Insider*, October 8, 2015. http://www. businessinsider.com/duckduckgo-ceo-its-a-myth-that-search-engines-need-to-track -you-2015-10.

Chapter 3: Behavioral Targeting

1. Jennifer Valentino-Devries, "What They Know about You," *Wall Street Journal*, July 31, 2010; Jennifer Valentino-Devries, "How to Avoid the Prying Eyes," *Wall Street Journal*, July 30, 2010; Amir Efrati, "'Like' Button Follows Web Users," *Wall Street Journal*, May 18, 2011; Julia Angwin and Jennifer Valentino-Devries, "Apple, Google Collect User Data," *Wall Street Journal*, April 22, 2011; Robert Lee Hotz, "The Really Smart Phone," *Wall Street Journal*, April 23, 2011; Jennifer Valentino-Devries and Emily Steel, "'Cookies' Cause Bitter Backlash," *Wall Street Journal*, September 19, 2010; Julia Angwin and Emily Steel, "Web's Hot New Commodity: Privacy," *Wall Street Journal*, February 28, 2011.

2. Harriet Taylor, "Privacy Will Hit Tipping Point in 2016," NBCNews.com, November 9, 2015.

3. Juniper Research, Digital Advertising: Online, Mobile & Wearables 2015–2019, Analyst Team, April 14, 2015.

4. "The End of Privacy" series, *Wall Street Journal*, 2012.

5. Ibid.

6. David G. Savage, "Supreme Court Case Pits Privacy Rights against Internet Data Brokers," *Los Angeles Times*, November 1, 2015. http://www.latimes.com/nation/la-na -supreme-court-data-privacy-20151102-story.html.

7. Steve Croft, "The Data Brokers," *60 Minutes*, March 9, 2014. http://www. cbsnews.com/news/data-brokers-selling-personal-information-60-minutes.

8. Emily Steel and Julia Angwin, "The Web's Cutting Edge, Anonymity in Name Only," *Wall Street Journal*, August 4, 2010.

9. Elizabeth Dwoskin, "Are Online Vendors Giving You Their Best Price? Maybe Not, Research Says," *Wall Street Journal*, October 23, 2014; Dana Mattioli, "On Orbitz, Mac Users Steered to Pricier Hotels," *Wall Street Journal*, August 23, 2012.

10. Mary Ann Russon, "Mac and Android Users Charged More on Shopping Sites Than iPhone and Windows Users," November 12, 2014. http://www.ibtimes .co.uk/look-out-you-might-be-charged-more-if-you-shop-online-using-mac-android -device-1474431.

11. Jessica E. Vascellaro, "Websites Rein in Tracking Tools," *Wall Street Journal*, November 9, 2010.

12. Jim Harper, "It's Modern Trade: Web Users Get as Much as They Give," *Wall Street Journal*, August 7, 2010.

13. Melanie Alnwick, "Internet Surfing Can Invade Privacy," MyFoxDC.com, May 26, 2010.

14. Efrati, "'Like' Button Follows Web Users."

15. "Global Mobile Statistics 2011: All Quality Mobile Marketing Research, Mobile Web Stats, Subscribers, Ad Revenue, Usage, Trends . . . ," MobiThinking.com, March 2011.

16. "Top Ten Uses for a Mobile Phone? Calls Come SIXTH!," *Daily Mail*, October 30, 2014. http://www.dailymail.co.uk/news/article-2815114/Top-ten-uses-mobile-phone -Calls-come-SIXTH-40-smartphone-users-say-manage-without-call-function-device .html.

17. Testimony on Behavioral Advertising: Industry Practices and Consumers' Expectations. Digital Marketing Speeches, Testimony to the House Committee on Energy and Commerce, Subcommittee on Commerce, Trade, and Consumer Protection, and the Subcommittee on Communications, Technology, and the Internet, for the Hearing on Behavioral Advertising: Industry Practices and Consumers' Expectations.

18. Jeremy Hsu, "Virtual Behavior Labs Discover What Gamers Want," LiveScience .com, February 16, 2011.

19. Anita Ramasastry, "Web Sites Change Prices Based on Customers' Habits," Special to CNN.com, June 24, 2005.

20. Derrick Harris, "Skylabs Says Video Analytics Are the Future," Gigaom.com, July 10, 2014.

Chapter 4: Self-Examination

1. John Seigenthaler, "A False Wikipedia 'Biography,'" *USA Today*, November 29, 2005.

2. Ibid.

3. Ibid.

4. Joel Stein, "Data Mining: How Companies Now Know Everything about You," *Time* (in cooperation with CNN), March 10, 2011.

5. Stein, "Data Mining."

6. DJ Patil, "Still the Sexiest Profession Alive," *Harvard Business Review*, November 21, 2013.

7. Interview with Lorraine Twohill, *McKinsey Quarterly*, February 2015.

8. Adam Tanner, "Here Are Some Companies Who Unmask Anonymous Web Visitors (and Why They Do It)," *Forbes*, July 1, 2013.

9. Kevin Poulsen, "The FBI Used the Web's Favorite Hacking Tool to Unmask Tor Users," *Wired*, December 16, 2014.

10. Poulsen, "FBI Used Hacking Tool."

11. Google Inside Search website, page on Algorithms, including questions and answers.

Chapter 5: Time to Get Dressed

1. CBS Interactive, Terms of Use for Internet Sites, Section 6 (User Submissions), Effective date October 16, 2014.

Chapter 6: Protecting Identity in a Crisis

1. Gemalto's 2014 Breach Level Index. http://www.gemalto.com/press/Pages/Gemalto-Releases-Findings-of-2014-Breach-Level-Index.aspx.
2. Tamara E. Holmes, "Credit Card Fraud and ID Theft Statistics," CreditCards.com September 16, 2015. http://www.nasdaq.com/article/credit-card-fraud-and-id-theft-statistics-cm520388.
3. Gemalto's 2014 Breach Level Index.
4. "What Is Identity Theft, and How Can It Impact You?" ID Alerts Canada Inc. http://www.idalerts.ca/about-identity-theft; Aleksandar Jevtic, "11 Countries with the Highest Rates of Identity Theft in the World," June 6, 2015. http://www.insidermonkey.com/blog/11-countries-with-the-highest-rates-of-identity-theft-in-the-world-351940.
5. "2015 Identity Fraud Report," Javelin Strategy & Research, March 2015.
6. "FTC Releases List of Top Consumer Complaints for 2010: Identity Theft Tops the List Again," U.S. Federal Trade Commission Press Release, March 8, 2011.
7. Symantec Press Release, Cupertino, CA, September 10, 2009.
8. Sara Gorman, "Identity Theft Twice as Likely in English-Speaking Countries: PayPal Trust and Safety Study Reveals That Online Fraud and Identity Theft Are Global Concerns," PayPal Press Release, October 21, 2008.
9. Caroline Davies, "Welcome to DarkMarket—Global One-Stop Shop for Cybercrime and Banking Fraud," *Guardian*, January 14, 2010.
10. Greg Aaron and Rod Rasmussen, "Global Phishing Survey 2H2014: Trends and Domain Name Use," APWG Industry Advisory, May 27, 2015. http://internetidentity.com/wp-content/uploads/2015/05/APWG_Global_Phishing_Report_2H_2014.pdf.
11. John E. Dunn, "The World's Biggest Data Breaches 2015—888 Incidents, 246 Million Records, Uncounted Misery," *Tech World*, September 9, 2015. http://www.techworld.com/picture-gallery/security/worlds-biggest-data-breaches-2015-888-incidents-246-million-records-uncounted-misery-3625117/#1; "VTech Jack: Data of 6.4M Kids Exposed," CNBC, December 2, 2014. http://www.cnbc.com/2015/12/02/vtech-hack-data-of-64m-kids-exposed.html.
12. Harriet Taylor, "Privacy Will Hit Tipping Point in 2016," NBCNews.com, November 9, 2015.
13. Juliana Gruenwald, "Google Agrees to Audits under FTC Settlement over Buzz," *National Journal*, March 30, 2011.
14. "The Ranking Digital Rights 2015 Corporate Accountability Index," November 3, 2015. https://rankingdigitalrights.org.
15. Alexis Madrigal, "Revealing the Man Behind @Mayoremanuel," *Atlantic Monthly*, February 28, 2011.
16. California Penal Code, Section 528.5.

17. "Factsheet on the 'Right to Be Forgotten' Ruling," C-131/12, European Commission. http://ec.europa.eu/justice/data-protection/files/factsheets/factsheet_data_protec tion_en.pdf.

Chapter 7: Branding Your Public Persona

1. Lauren Indvik, "University to Provide Online Reputation Management to Graduates," Mashable.com, May 5, 2010.
2. Jennifer Van Grove, "How Job Seekers Are Using Social Media for Real Results," Mashable.com, March 8, 2010.
3. Jennifer Fernicola Ronay, "Facebook Lawsuit: Mom, Son Say Pranksters Set Up Racist, Sexual Profile," ChicagoNow.com, September 24, 2009.
4. "Top 15 Most Popular Social Networking Sites, November 2015." http://www .ebizmba.com/articles/social-networking-websites.
5. Jennifer Van Grove, "How Job Seekers Are Using Social Media for Real Results," Mashable.com, March 8, 2010.
6. Josh Grossberg, "Courtney Love in Trouble for Tweeting," MSNBC.com, May 27, 2011.

Chapter 8: Your Right to Be Forgotten and to Complain Online

1. Caitlin Dewey, "How the 'Right to Be Forgotten' Could Take Over the American Internet, Too," *Washington Post*, August 4, 2015.
2. Jeffrey Toobin, "The Solace of Oblivion," *The New Yorker*, September 29, 2014.
3. *Kashian v. Harriman*, 98 Cal. App. 4th 892, 120 Cal. Rptr. 2d 576.
4. 128 Cal. App. 4th 1569, 27 Cal. Rptr. 3d 863, California Court of Appeal, First District, Division 4, 2005.
5. Ibid.
6. Josh Harkinson, "Yelp Is Pushing a Law to Shield Its Reviewers from Defamation Suits," *Mother Jones*, July 20, 2015.

Chapter 9: Dress for Career Success

1. Businessdictionary.com: The ratio of value to cost of adding one more network user grows disproportionately as the network grows larger.

Chapter 10: Don't Forget the Kids

1. Andrea Peterson, "Google Is Tracking Students as It Sells More Products to Schools, Privacy Advocates Warn," *Washington Post*, December 28, 2015. https://www .washingtonpost.com/news/the-switch/wp/2015/12/28/google-is-tracking-students-as -it-sells-more-products-to-schools-privacy-advocates-warn.

2. Maeve Duggan, Amanda Lenhart, Cliff Lampe, and Nicole B. Ellison, "Parents and Social Media: Concerns about Children, Social Media, and Technology Use," Pew Research Center, July 16, 2015. http://www.pewinternet.org/2015/07/16/concerns -about-children-social-media-and-technology-use.

3. Amanda Lenhart, "Teens, Social Media & Technology Overview 2015," April 9, 2015. http://www.pewinternet.org/2015/04/09/teens-social-media-technology-2015.

4. Amanda Lenhart, Rich Ling, Scott Campbell, and Kristen Purcell, "Teens and Mobile Phones," Pew Internet and American Life Project, April 20, 2010.

5. "U.S. Teen Mobile Report: Calling Yesterday, Texting Today, Using Apps Tomorrow," Nielsen Company, October 14, 2010.

6. "Digital Diaries," AVG, January 19, 2010.

7. Tamar Lewin, "If Your Kids Are Awake, They're Probably Online," *New York Times*, January 20, 2010.

8. Robert Hackett, "Here's How Teens Really Use Their Phones," *Fortune*, May 27, 2015. http://fortune.com/2015/05/27/teens-phones-mary-meeker.

9. Chris Matyszczyk, "Zuckerberg: I Know That People Don't Want Privacy," CNET.com, January 10, 2010.

10. Julia Angwin, "The Web's New Gold Mine: Your Secrets," *Wall Street Journal*, July 30, 2010.

11. AOL and the Nielsen Company surveyed more than one thousand adults and five hundred teens ranging in age from thirteen to seventeen via an e-mail survey (August 2010).

12. Lorenzo Franceschi-Bicchierai, "One of the Largest Hacks Yet Exposes Data on Hundreds of Thousands of Kids," November 27, 2015. http://motherboard.vice.com/ read/one-of-the-largest-hacks-yet-exposes-data-on-hundreds-of-thousands-of-kids.

13. National Crime Prevention Council, Cyberbullying page, http://www.ncpc.org/ topics/cyberbullying.

14. Jaime Sarrio, "Tennessee Teen Expelled for Facebook Posting," *The Tennessean*, *USA Today*, January 28, 2010.

15. Eric Frazier, "Facebook Post Costs Waitress Her Job," *Charlotte Observer*, May 17, 2010.

16. Frazier, "Facebook Post Costs Waitress."

17. Michelle Esteban, "3 Arrested after Teen's Nude Photo Sweeps through Schools," KOMO News, January 28, 2010.

18. Alex Sanz, "FBI: Social Networking Sites a Favorite Target of Child Predators," KHOU Houston, June 9, 2010.

19. "The Kids Are Alright: A Study of the Privacy Habits of Parents and Their Teens on Social Networks," *TRUSTe and Lightspeed Research*, October 2010.

20. Jason Mick, "Facebook Faker Stands Accused of Blackmailing 31 Males for Sex," *Daily Tech*, February 5, 2009.

21. Dinesh Ramde, "Anthony Stancl, 19, Gets 15 Years for Facebook Sex Scam," *Huffington Post Tech*, February 24, 2010; Charles Wilson, "Online 'Sextortion' of Teens on Rise, Feds Say," Associated Press, August 14, 2010.

22. Jane Sims, "Child Porn Stash 'Largest': Cybercrime: U.S. Homeland Security Tipped off London Police as They Bust Major Link in a North American Ring," *London Free Press*, December 17, 2010.

23. Sally Weale, "Teens' Night-time Use of Social Media 'Risks Harming Mental Health,'" *Guardian*, September 10, 2015.

24. "Young Mobile Users Drive UK Social Media Usage, Social Networking Popular among Teens and Millennials," *eMarketer*, October 29, 2015. http://www.emarketer .com/Article/Young-Mobile-Users-Drive-UK-Social-Media-Usage/1013163.

25. Amanda Lenhart, "Teens, Social Media & Technology Overview 2015," April 9, 2015. http://www.pewinternet.org/2015/04/09/teens-social-media-technology-2015.

26. "U.S. Teen Mobile Report."

27. SwiftKey Emoji Report, April 21, 2015. http://www.scribd.com/doc/262594751/ SwiftKey-Emoji-Report.

28. "Teenage Sexting Statistics," *GuardChild*, April 14, 2014.

29. "New Jersey Principal Wants to Keep Middle School Kids Off Facebook—Do You Agree?" *ABC World News with Diane Sawyer*, April 29, 2010.

30. Jason Kessler, "Principal to Parents: Take Kids off Facebook," CNN, April 30, 2010.

31. Andrea Petrou, "Kids Give Their Parents the Runaround Online," TecheYe.net, April 19, 2011.

32. "Reach of Leading Social Media and Networking Sites Used by Teenagers and Young Adults in the United States as of February 2015." http://www.statista.com/statis tics/199242/social-media-and-networking-sites-used-by-us-teenagers.

33. "Social Networking Sites—Quick Facts," OnGuardOnline.gov.

34. Jan Hoffman, "As Bullies Go Digital, Parents Play Catch-Up," *New York Times*, December 4, 2010.

35. Ibid.

36. "Location and Privacy: Where Are We Headed" and "Online Reputation in a Connected World," Cross-Tab, 2011, and Microsoft, 2011.

37. Natasha Singer, "They Loved Your G.P.A. Then They Saw Your Tweets," *New York Times*, November 9, 2013. http://www.nytimes.com/2013/11/10/business/ they-loved-your-gpa-then-they-saw-your-tweets.html.

38. John Hechinger, "College Applicants, Beware: Your Facebook Page Is Showing," *Wall Street Journal*, September 18, 2008.

39. Ibid.

40. Marc Beja, "How Students, Professors, and Colleges Are, and Should Be, Using Social Media," *Chronicle of Higher Education*, August 24, 2009.

41. "Can What You Post on Facebook Prevent You from Getting into College?" Reporter: Yoni, Inside the Admissions Office, An Online Forum, *Wall Street Journal on Campus* (video).

Chapter 11: Turning Off the Lights

1. Bianca Bosker, "Eric Schmidt on Privacy: Google CEO Says Anonymity Online Is 'Dangerous,'" *Huffington Post*, October 10, 2010.

2. Kelli B. Grant, "Identity Theft Victims: You Might Know the Culprit," CNBC, July 21, 2015.

3. "BBB Warns of Phishing Email Received from Epsilon Data Breach," Better Business Bureau, April 7, 2011; Miguel Helft, "After Breach, Companies Warn of

E-Mail Fraud," *New York Times*, April 4, 2011; BankOrion's Customer Alert, Epsilon Fraud Watch at http://bankorion.com/epsilon-alert.php; "Hotels: Ritz-Carlton Customers' Data Stolen in Hack Attack," Hospitality, Chief Officers' Network, April 7, 2011.

4. Jennifer Steinhauer, "Verdict in MySpace Suicide Case," *New York Times*, November 26, 2008.

5. "Privacy Please! U.S. Smartphone App Users Concerned with Privacy When It Comes to Location," Nielsen Wire, Nielsen Company, April 2011.

6. "Who's Watching You?" Opera Press release, January 28, 2011.

7. Sarah Kessler, "What Is Data Privacy Day?" Mashable.com, January 28, 2011.

8. "'Happening Now': Facebook Cooks Up Another Privacy Breach," Facecrooks.com, June 11, 2011.

9. Robin Wauters, "Third-Party Twitter Apps Can Access Your Private Messages without Authorization," *TechCrunch*, June 10, 2011.

10. Erik Kirschbaum and Jon Boyle, "Facebook Blunder Leads Crowd to Teen's Birthday," Reuters, June 5, 2011.

11. "On the Internet, Nobody Knows You're a Dog," Cartoon Bank, *New Yorker* Cartoon & Cover Prints, *New Yorker*, July 5, 1993.

12. "Cyber Norton Report: The Human Impact." http://us.norton.com/theme.jsp?themeid=cybercrime_report.

13. "25% of Internet Users Use a Fake Name," *Our Daily*, September 12, 2010.

14. Amy Lee, "U.S. Military Launches Spy Operations Using Fake Online Identities," *Huffington Post*, March 17, 2011; Brad Stone and Matt Richtel, "The Hand That Controls the Sock Puppet Could Get Slapped," *New York Times*, July 16, 2007.

15. Jolie O'Dell, "Woman Catches Husband in Fake Murder Plot with Fake Facebook Profile," Mashable.com, June 10, 2011.

16. David Kesmodel and John R. Wilke, "Whole Foods Is Hot, Wild Oats a Dud—So Said 'Rahodeb,'" *Wall Street Journal*, July 12, 2007.

17. A. Jeff Ifrah and Steven Eichorn, "Commentary: Banned from the Internet," *National Law Journal*, October 13, 2010.

INDEX

ABOUT THE AUTHORS

Ted Claypoole is an attorney practicing in data management, software service agreements, and the law of the Internet. Ted leads his law firm's IP transaction team and its privacy and cybersecurity team. He is also chair of the Cyberspace Law Committee for the Business Law Section of the American Bar Association. He speaks and writes regularly about the intersection of law and technology.

Theresa Payton currently runs Fortalice Solutions, LLC, a security consulting firm, and is the cofounder of Dark3, LLC, a cybersecurity product company. Previously, she was the chief information officer at the White House, spearheading technology and security efforts at the highest level of the U.S. government. Before government service, she had a distinguished career as a senior executive in the financial services industry at various banks, giving her a valuable perspective in terms of protecting information and insights into the tradecraft of fraudsters and other criminal actors. Theresa's experience in both government and banking makes her a leading authority and highly sought-after expert on cyber issues. For the latest consumer and kid-safety advice, you can watch her segments on WBTV CBS News (in Charlotte, NC), *Today*'s "Hacking of America" series by Tom Costello, and *Good Morning America*.